CONTENTS

S. Hrg. 109–619

BIO-SECURITY COORDINATION

HEARING

BEFORE THE

SUBCOMMITTEE ON RESEARCH, NUTRITION, AND GENERAL LEGISLATION

OF THE

COMMITTEE ON AGRICULTURE, NUTRITION, AND FORESTRY UNITED STATES SENATE

ONE HUNDRED NINTH CONGRESS

SECOND SESSION

January 9, 2006

Printed for the use of the
Committee on Agriculture, Nutrition, and Forestry

Available via the World Wide Web: http://www.agriculture.senate.gov

U.S. GOVERNMENT PRINTING OFFICE
30–234 PDF WASHINGTON : 2006

For sale by the Superintendent of Documents, U.S. Government Printing Office
Internet: bookstore.gpo.gov Phone: toll free (866) 512–1800; DC area (202) 512–1800
Fax: (202) 512–2250 Mail: Stop SSOP, Washington, DC 20402–0001

COMMITTEE ON AGRICULTURE, NUTRITION, AND FORESTRY

SAXBY CHAMBLISS, Georgia, *Chairman*

RICHARD G. LUGAR, Indiana
THAD COCHRAN, Mississippi
MITCH McCONNELL, Kentucky
PAT ROBERTS, Kansas
JAMES M. TALENT, Missouri
CRAIG THOMAS, Wyoming
RICK SANTORUM, Pennsylvania
NORM COLEMAN, Minnesota
MICHEAL D. CRAPO, Idaho
CHARLES E. GRASSLEY, Iowa

TOM HARKIN, Iowa
PATRICK J. LEAHY, Vermont
KENT CONRAD, North Dakota
MAX BAUCUS, Montana
BLANCHE L. LINCOLN, Arkansas
DEBBIE A. STABENOW, Michigan
E. BENJAMIN NELSON, Nebraska
MARK DAYTON, Minnesota
KEN SALAZAR, Colorado

MARTHA SCOTT POINDEXTER, *Majority Staff Director*
DAVID L. JOHNSON, *Majority Chief Counsel*
STEVEN MEEKS, *Majority Legislative Director*
ROBERT E. STURM, *Chief Clerk*
MARK HALVERSON, *Minority Staff Director*

BIO-SECURITY COORDINATION

MONDAY, JANUARY 9, 2006

U.S. SENATE,
SUBCOMMITTEE ON RESEARCH, NUTRITION, AND GENERAL
LEGISLATION OF THE COMMITTEE ON AGRICULTURE,
NUTRITION AND FORESTRY,
Washington, DC.

The subcommittee met, pursuant to notice, at 1:15 p.m. in the Farm Show Complex, 2300 North Cameron Street, Harrisburg, Pennsylvania, Hon. Rick Santorum, chairman of the subcommittee, presiding.

Present or submitting a statement: Senator Santorum.

OPENING STATEMENT OF RICK SANTORUMA, A U.S. SENATOR FROM PENNSYLVANIA, CHAIRMAN, SUBCOMMITTE ON RESEARCH, NUTRITION, AND GENERAL LEGISLATION, COMMITTEE ON AGRICULTURE, NUTRITION, AND FORESTRY

Senator SANTORUM. Good afternoon. Let me thank everybody for attending here this afternoon, and I particularly want to thank the panelists who are here with us today. We have got three great panels to hear about the state of agriculture in the area of biosecurity. It is a very important time to be talking about that. There has certainly been a lot reported in the press about concerns with terrorism, as well as concerns with just the natural infections that come and that could affect our food supply, and to have a discussion as Chairman of the Senate Subcommittee on Research and Nutrition—this is an area of jurisdiction of the Agriculture Committee. It is something that we have had hearings on in Washington, and we wanted to take the opportunity, with the focal point on agriculture here in Pennsylvania with the Farm Show, to talk about its impact on Pennsylvania and the impact on agriculture in this State and on our economy, generally.

I know our good Secretary, Secretary Wolff, is here, and certainly, as I often do sing the praises of Pennsylvania agriculture and talk about its importance to the economy of this Commonwealth, it is our No. 1 industry and we are very proud of that. Anybody walking into the halls today can see it and say it is a very important industry and one that has a lot of folks participating in it.

It has been a great interest of mine. I am very proud to say that I was the first United States Senator on the Agriculture Committee from Pennsylvania since—almost 100 years ago was the last Pennsylvania Senator on the Agriculture Committee. I always like to say that you can tell our lack of representation has had its telltale signs in agriculture policy, which has been written principally by folks representing the Midwest and the Southern States, and it is important to have folks with a different point of view who represent States like Pennsylvania, as the Secretary will tell you, which is as diverse a State agriculturally as maybe any State in the country.

We have great diversity, but as a result of that, there is not—like there are from other States—folks who are locked into particular crops to make sure that those particular crops are taken care of, as we see in various States around the country. Agriculture policy has been drafted to sort of take care of some of these Corn States, Wheat States, Cotton States, Rice States, as opposed to us, which is sort of an Everything State, if you will. So having that voice on the committee is important and having the cooperation that I certainly have gotten from the producer community here in Pennsylvania has been vital to that. We have accomplished a lot of things for Northeastern agriculture over the past several years, not the least of which is just recently where we were able, although it is not quite done yet, to keep the MILC payments in place. This was, I must tell you, a hard-fought battle, and I thank the Secretary. I know that he went to work in talking to other Secretaries of Agriculture in the region.

[Laughter.]

Senator SANTORUM. We were able to take a program that was scheduled to expire last year, and as a result of a lot of hard work and teamwork, we were able to put that extension in place. It has not yet passed. It has passed the Senate in a conference report, and we anticipate the House passing it relatively soon this year. So those payments will be available hopefully very, very soon, but that is a major success. It sets us up for a farm bill where that program is on the table for discussion, for the dairy program.

I will tell you that I have worked in a lot of legislative areas in the past, in a whole variety of areas in agriculture and outside of agriculture, and I cannot think of any tougher area to legislate than dairy policy. It is as regional and as specific to a region of the State as any policy, and we are going to have a battle, I'm just warning everybody, to be able to maintain this program. The MILC program, as Senator Kohl and I were the authors of—Senator Kohl from Wisconsin—we were able to put a team together and get that done, but I will tell you that there are folks from the West and the South in particular who would like to see that program go away. As you know, that program is beneficial to smaller farms, although every dairy farm benefits from it. It is beneficial to smaller farms because it is a limit on—it only pays up to a certain limit of production and so the bigger farmers do not care as much about it. They care about the price support program, which is based on Class III milk as opposed to MILC payments, which are more beneficial to those of us who produce a lot of fluid milk. So it is going to be that kind of battle that we are going to be engaged in, not just in milk, but in a whole variety of issues coming up with the farm bill.

One of the issues that—I know the Secretary has been here, Chuck Conner, who is here representing the Secretary today, was up at Penn State last year to do a listening session on the farm bill. I have been doing the same thing, traveling the State, trying to get feedback. It is going to be a monumental year. We have got a big issue coming up in 2007 next year with the farm bill coming forward, and I certainly hope that I am in a position as the No. 3 leader in the Senate, hopefully No. 2 by next year, to be able to influence that policy and working to make sure that Pennsylvania's agriculture interests, whether the work that we have done on specialty crops, the work that we have done on crop insurance, a whole host of other things, farmland protection, all of those things we were able to get in previous farm bills as a member of the Agriculture Committee and as a member of leadership, and I just want to assure you that we will certainly do all we can to make sure that interest of Northeastern agriculture—and when I say Northeastern, I don't mean northeastern Pennsylvania, I mean the Northeastern U.S.—are represented on a committee that does not have too many of us. Senator Leahy and I are the only two, and certainly he and I do not necessarily see eye to eye on a lot of things, but we do on several agriculture programs, milk being one of them. So I just wanted to give you a little backdrop of why we are holding these hearings.

One of the other issues we are going to be dealing with are issues of biosecurity. We have done some things that are important in this bill. In one of the bills that we passed at the end of the year, we were able to pass several billion dollars in money for a Presidential initiative on avian influenza, which is not as much, candidly, as we wanted. We wanted to pass $7 billion. I think we got three-point—yes, three-point-something. I do not remember the exact number, roughly half of what the President suggested. We will probably go back and have to work to get the rest, but it is vitally important and we will hear discussions today about how we deal with this potential pandemic for us, as well as the impact that it will have on agriculture and our food security. So those are issues that we are dealing with now, but we will be dealing with on a more comprehensive basis when we look at the farm bill.

Let me stop and turn it over to the folks who are the experts in agriculture. I just try to listen to the experts and do my best to try to represent the folks here in Pennsylvania, but we have three gentleman who are here representing—two from the Federal Government, one who is well-known to the folks here in Pennsylvania. As I mentioned several times, Secretary Wolff is here, but I also want to introduce Chuck Conner, who is the Deputy Secretary of Agriculture.

One of the things I found out, just for those who do not know, the shorter the title you have in a department, the more important you are, just so you understand that. So when you are talking to the Under Deputy Secretary in Charge of, you are way down the list. If you are the Secretary, you are the big guy. If you the Deputy Secretary, you are the second-biggest guy. Then you get into Under Secretary, but Deputy Secretary, that is a good title to have. That means you are very important, and Chuck is here. I had a chance to know and work with Chuck as he was the lead staff person, the staff director, on the Senate Agriculture Committee. It is good to have you here, Chuck. Thank you for being here.

Alex Azar, who is the Deputy Secretary of Health and Human Services—another one of those short titles, which means that Alex is a person with great responsibility at the Department of Health and Human Services. I want to thank both of those gentlemen for attending, as well as you, Mr. Secretary. Again, thank you for your attendance.

Chuck, why don't you start off our testimony.

STATEMENT OF HON. CHARLES CONNER, DEPUTY SECRETARY OF AGRICULTURE, USDA, WASHINGTON, DC

Mr. CONNER. Well, Chairman Santorum, thank you for holding this hearing today—is my mike on?

Senator SANTORUM. I do not think so. Try the top. Look at the top. See if it is on.

Mr. CONNER. Is that on now, Mr. Chairman?

Senator SANTORUM. Yes, it is on now.

Mr. CONNER. OK. Thank you. Chairman Santorum, again thank you very much for holding this hearing. On behalf of the Department of Agriculture and the Bush administration, I just want to thank you for your leadership on this issue. I also thank you, as I have observed for many, many years your advocacy on behalf of Northeast agriculture as it benefits the State of Pennsylvania, and I commend you for that.

As you know, Mr. Chairman, biosecurity is an issue that we take very, very seriously at the USDA, and we appreciate your efforts to highlight its importance to our Nation. As you are also aware, I testified before the full Agriculture Committee in July of 2005, on food and agriculture security issues. My testimony today should provide an important update on that information, and it also should include more discussion of the Federal-State collaborative research efforts, and, of course, Mr. Chairman, I will submit my full statement for the record for you and summarize those remarks.

Senator SANTORUM. Thank you.

Mr. CONNER. USDA considers food and agriculture biosecurity issues essential to its mission. The success of USDA's efforts is dependent upon a coordinated work of a broad range of Federal, State, local and private sector partners. Let me begin by putting the importance of agriculture into some perspective for you. Agricultural exports, as you know, accounted for $62.4 billion in fiscal year 2005, and we are expected to reach record high dollar values of $64 billion in 2006, with agricultural trade, of course, being a very positive addition to our balance of payments and trade. Agricultural imports, as well, provide significant value, over $57 billion in 2005. Our Nation's food and agricultural system contributes approximately $1.24 trillion, over 12 percent, to our gross domestic product, and we are directly responsible for employing 17 percent of our Nation's workforce. We are an important sector indeed, certainly important to the State of Pennsylvania.

We face many challenges in protecting this important infrastructure. The agricultural industry is particularly concerned about security because diseases and pathogens, whether they occur naturally or are unintentionally introduced or are intentionally delivered, do not respect geopolitical borders. While the interconnected nature of the global food system contributes to our own economic strength by improving production and marketing efficiency and providing timely responses to our consumers, our integrated system is also challenged in the event of an attack or natural disease outbreak. Products in this country move very, very quickly from State to State and from Nation to Nation. A pest, a disease or another agent can spread just as quickly.

Since September 11, 2001, USDA has made great strides in expanding our mission to better address food and agricultural security. We believe that the threat to agriculture is indeed very real. The department has been working closely with its Federal, State and local government partners, as well as with industry stakeholders, to address these concerns using a sector-wide strategy based upon White House guidance. My written testimony highlights the advances that the department is making to implement Homeland Security Presidential Directive 7, Critical Infrastructure Identification, Prioritization and Protection, and, of course, Homeland Security Presidential Directive 9, Defense of the United States Agriculture and Food.

USDA has worked in coordination with the Department of Homeland Security and are active partners in the Department of Health and Human Services Food and Drug Administration to ensure that we develop a coordinated approach for protecting the food and agricultural sector. Central to this effort, Mr. Chairman, is our collaboration with the private sector for infrastructure protection and to create an overarching framework and unique plans for protecting key assets and resources.

We have worked with Federal, State, local and private sector participants to establish avenues for sharing sensitive information, new policies, best practices and, of course, vulnerability assessments. We have formed two counsels, one for government, one for the private sector. The Food and Agriculture Government Coordinating Council is led jointly by the Department of Homeland Security, USDA and FDA, and includes Federal, State and local officials. The private sector council, called the Food and Agriculture Sector Coordination Council, represents a table-to-farm continuum. The two counsels meet quarterly and regularly hold individual and joint calls to discuss issues of mutual interest, such as sector vulnerability assessments and Federal research and development plans, in which, of course, you have a strong interest.

Preparation, research, surveillance and response are all integral parts to our ability to respond to biosecurity threats, whether they are related to intentional acts of terrorism or the unintentional introduction of agents, pests or disease. The Federal Government, State and local governments and private stakeholders must be partners in those efforts if we are to fully achieve security. I am pleased to report to you today that we are indeed partners in this effort and are making great headway. Our integrated Federal-State system of collecting information and samples to detect diseases, utilizing state-of-the-art laboratory networks to test samples and jointly analyzing data, shows the benefits of our collaboration. Through this system, we are able to anticipate new and emerging threats and to quickly eliminate and contain already identified existing threats. I am confident that all of us here today are united in our belief that biosecurity for food and agriculture is a paramount issue that must be addressed with sound science and sound policy decisions. The department looks forward very much to continuing to work with this committee in a continuing effort to develop programs and initiatives to help enhance the security of our Nation's agriculture and food systems through collaborative efforts between Federal and State institutions.

Mr. Chairman, that concludes my oral remarks and at the appropriate time I would be happy to take any questions that you may have for the Department of Agriculture.

[The prepared statement of Mr. Conner can be viewed in the appendix on page 36.]

Senator SANTORUM. Thank you, Mr. Conner.

Mr. Azar?

STATEMENT OF ALEX M. AZAR, II, ESQUIRE, DEPUTY SECRETARY, DEPARTMENT OF HEALTH AND HUMAN SERVICES, WASHINGTON, D.C.

Mr. AZAR. Good afternoon, Chairman Santorum. I am Alex Azar, Deputy Secretary of the United States Department of Health and Human Services, of which the Food and Drug Administration is a part. Thank you very much for holding these hearings on these critical issues and thank you very much for your important and ongoing leadership in the agriculture and food safety areas, areas of fundamental importance for us at HHS. I have a more complete statement that I have submitted for the record, but would like to just summarize some of the points.

I note that we are on the grounds of the Pennsylvania Farm Show Complex, which is the largest agricultural event in America, and this is, of course, a fit setting for today's hearings, and I appreciate this opportunity for us to be able to discuss our food counterterrorism activities. As a Johnstown, Pennsylvania native, I am also very pleased to be back in the State of my birth.

Before I address the main subject of my testimony, I am pleased to tell you that last week Secretary Leavitt announced $100 million in emergency contingency fund to States, Tribes and Territories that will help families in need pay their heating and energy bills this winter. We are also making 100 percent of the State's remaining block grant funds available, which amounts to $633 million, for a combined total of $733 million in low-income heating and cooling assistance. Pennsylvania's share of that release was almost $7.7 million.

On the subject of today's testimony, under the President's National Response Plan, HHS is responsible for leading the Federal public health efforts to ensure an integrated and focused national effort to anticipate and respond to emerging biological and other weapons threats. HHS is also the principal Federal agency responsible for coordinating all Federal level assets activated to support and augment the State and local medical and public health response to any mass casualty events. Within HHS, the Centers for Disease Control and Prevention, the Health Resources and Services Administration and the National Institutes of Health play key roles in our overall counterterrorism activities. For example, CDC operates HHS's Strategic National Stockpile, which contains large quantity of medicines and medical supplies for use in the event of a public health emergency. NIH is taking the lead role in the development of medical countermeasures to a biological attack.

The Food and Drug Administration in HHS is the Federal agency that regulates everything we eat except meat, poultry and processed egg products, which are regulated by our partners at USDA. FDA's responsibility also extends to live animals and animal feed, and FDA is responsible for ensuring that human drugs, human biological products, medical devices and radiological products, as well as veterinary drugs, are safe and effective and that cosmetics are safe.

In the past few years, HHS has done a great deal to enhance the safety and security of the food supply of the United States. Food safety and food defense continue to be top priorities for the administration. A terrorist attack on the food supply could have both severe public health and economic consequences while damaging the public's confidence in the food we eat. The changes in food safety and defense that we have been implementing in the last few years are the most fundamental enhancements in our food safety and defense activities in many years. FDA has worked with food safety agencies, as well as with law enforcement and intelligence gathering agencies and with industry, to significantly strengthen the Nation's food safety system across the entire distribution chain, from farm to table, to better protect our food supply against deliberate and accidental threats. This cooperation has resulted in greater awareness of vulnerabilities, the creation of more effective prevention programs, new surveillance systems and faster outbreak response capabilities.

One byproduct of these efforts, in collaboration between the Federal, State and local governments, as well as the private sector, is guidance that we have been able to publish for industry, a copy of which I have today, which is the guidance that the FDA issued for dairy farmers and dairy producers to ensure the safety of America's dairy products. FDA is working closely with the Department of Homeland Security and other Federal agencies to implement the President's Homeland Security Presidential Directives, as Secretary Conner said. The President has issued HSPDs 7, 8 and 9, which identify critical infrastructures, improve response planning and establish a national policy to defend the agriculture and food systems against terrorist attacks, major disasters and other emergencies. Over the past 3 years, FDA and HHS have also been implementing the Public Health Security and Bioterrorism Preparedness Act of 2002. The Bioterrorism Act provided the Secretary of HHS with significant new authorities to protect the Nation's food supply against the threat of intentional contamination and other food-related emergencies. These authorities improve our ability to act quickly in responding to a threatened attack, as well as other food-related emergencies.

I would like to mention just a few of those provisions. Section 305 of the Bioterrorism Act requires registration of foreign and domestic food facilities that manufacture, process, pack or hold food for consumption by humans or animals in the U.S. With this provision, FDA has for the first time ever a roster of foreign and domestic food facilities that provide food for American consumers. In the event of an emergency, the registration information will help FDA quickly identify, locate and notify the facilities that may be affected.

Section 307 requires the submission to FDA of prior notice of feed, including animal feed, that is offered for import into the United States. This advance information allows the FDA, working closely with the Customs and Border Patrol, to more effectively target inspections at the border to ensure the safety of imported foods before they move into the United States.

Section 306 authorizes FDA to have access to certain records when the agency has a reasonable belief that an article of food is adulterated and presents a threat of serious adverse health consequences or death to humans or animals. This enhances FDA's ability to track and contain foods that pose a threat to American consumers from accidental or deliberate contamination of food.

I would like to briefly mention a few other program activities. FDA has issued guidance on the security measures the food industry may take to minimize the risk that food will be subject to tampering or other malicious criminal or terrorist actions. To increase laboratory research capacity, FDA has worked closely in collaboration with FSIS at USDA to establish the Food Emergency Response Network, to include a substantial number of laboratories capable of analyzing foods for agents of concern.

We have also conducted extensive scientific vulnerability assessments of different categories of food, determining the most serious risks of intentional contamination with different biological or chemical agents during various stages of food production and distribution. We also have an ambitious research agenda to address potential terrorist threats. I would also be remiss, Mr. Chairman, if I did not mention the efforts underway in the administration and at HHS with regard to avian influenza. Secretary Mike Leavitt has made influenza pandemic planning and preparedness a top priority, and we thank you very much for your leadership in securing through Congress the passage of the first year's funding, the full funding for the first year, of the President's proposal for pandemic avian influenza preparedness, and I look forward to working with you together to secure the funding for the following 2 years so that we can complete the ambitious agenda we have laid out to get this country better prepared for a possible pandemic influenza outbreak.

Through the Influenza Preparedness Task Force that Secretary Leavitt has chartered, the FDA and other agencies within HHS are working together to prepare the U.S. for this potential threat to the health of our Nation. We are also working closely with other Federal, State, local and private sector entities to ensure close collaboration. Because pandemic influenza viruses will emerge in part or wholly from among animal influenza viruses, such as birds, it is critical for human and animal health authorities to closely coordinate their activities, such as surveillance, to share relevant information as quickly and transparently as possible. In that connection, HHS collaborates with USDA and the Department of Interior in critical partnerships for domestic preparedness for a possible avian influenza outbreak here in the United States. Early detection will be the key to allowing the United States to having the most up-to-date and reliable information that will help us save animal and human lives.

In conclusion, HHS is making significant progress in its ability to ensure the safety of the food supply due to the enhancements being made by FDA and other agencies and due to the close coordination between Federal and State food safety, public health, law enforcement and intelligence-gathering agencies. The U.S. food safety and defense system is stronger than ever before. Although we are better prepared than ever before, we are continuously working to improve our ability to prevent, detect and respond to terrorist threats.

Thank you very much for this opportunity to discuss our counterterrorism activities with you to protect the American food supply, and I would be pleased to respond to any of your questions at the appropriate time.

[The prepared statement of Mr. Azar can be viewd in the appendix on page 65.]

Senator SANTORUM. Thank you, Mr. Azar.

Mr. Secretary, thank you for being here.

STATEMENT OF DENNIS WOLFF, SECRETARY, PENNSYLVANIA DEPARTMENT OF AGRICULTURE, HARRISBURG, PENNSYLVANIA

Mr. WOLFF. Thank you, Senator. On behalf of the Pennsylvania Department of Agriculture, as well as the entire agriculture community in Pennsylvania, we want to thank you for choosing the largest indoor livestock exposition in the United States, the 90th Annual Pennsylvania Farm Show, as your venue for these hearings today. As mentioned earlier, agriculture is such a huge economic driver here in the Commonwealth of Pennsylvania, representing around $4.5 billion of farm gate revenue, and when you do the multiplier on that it has an impact of over $40 billion in Pennsylvania. It also represents one out of every seven jobs here in the Commonwealth.

Agriculture in Pennsylvania is very diverse. We have everything from Christmas tree farms to mushroom growers, to the largest growing hardwood industry in Pennsylvania, as well as we have a huge livestock component, and therein lies some of our concerns as it relates to bioterrorism and the concerns as to why we need a good system in place here in Pennsylvania.

Under Governor Rendell's direction, the Pennsylvania Department of Agriculture in 2003 was included as part of the Homeland Security and Domestic Preparedness Assessment and Strategic Planning Program. We were only one of six States in the United States that included agriculture in this plan and this program. This truly reflects the Commonwealth's commitment to agriculture biosecurity prevention, as well as response and recovery programs that we feel we need in place here in Pennsylvania.

Pennsylvania has invested in and implemented several new programs to improve our biosecurity systems. The funds that were used from PEMA and USDA were used to update our information system between the two laboratories here in Pennsylvania. Also, it was used to update our premise data base so that all the farms in Pennsylvania will be listed on a GIS map so that we can effectively and expediently carry out any actions we need to do in case of an outbreak within our livestock industry here in Pennsylvania.

We also used funds from Health and CDC to rebuild our food safety inspection system. We call it the Garrison System. It is real-time information. It is a good, consolidated system throughout Pennsylvania that really helps us do a much better job in working with all the restaurants in Pennsylvania that we inspect, as well as the food processing centers in Pennsylvania, which makes sure that everything is not only, in fact, being properly handled, but if there is an outbreak, it is an immediate flag so that we know that we have got a problem, something outside of the ordinary challenges that our food safety inspectors may need to deal with.

We have also developed the PASART program. This came from funds from Homeland Security and PEMA once again. It is to train local volunteers to develop a local and State plan as it relates to animal emergencies, and, in fact, Joel Hersh, who is our Executive Director, is working with USDA to help develop a national response plan with a similar type of mission.

We recently conducted a tabletop exercise for an avian influenza outbreak, and that was with our partners at Penn State, the University of Pennsylvania, the Department of Health, PEMA, the FBI, USDA and PennAg industries, so we had all our stakeholders at the table to look up both our strengths and our weaknesses so that we can do a better job in terms of responding to an outbreak if we had one here.

Governor Rendell recently approved construction for a BSL–3 lab here at the Harrisburg laboratory, which will give us the capabilities of handling samples that are more sensitive and are also—certainly in helping protect our staff that works with those samples. So what else do we need? We still have some needs here in Pennsylvania to make us even a better State in terms of being prepared. We feel that our plant industry is something that really can be upgraded in terms of dealing with a timely response to issues such as what rust, corn leaf blight or plum pox.

We think that the BSL–3 laboratory capability should be included in our other two diagnostic labs. Here in Pennsylvania, we have three laboratories. We have one here in Harrisburg, one at the University of Pennsylvania and one at Penn State University. Only one of those three labs is going to have the BSL–3 capacity. We think we need funding for more personal protective gear as it deals with issues such as an outbreak of H5N1 in our poultry industry.

We also think that we should have more funds in terms of research and development as it relates, in particular, to agriculture and domestic preparedness. I hope this gives you some idea of some of the programs that we have here in Pennsylvania, as well as some of our additional needs. I would like to thank you for being here today, and we hope you enjoyed the 90th Farm Show. I will also be submitting written testimony and be glad to answer questions when the time is appropriate.

[The prepared statement of Secretary Wolff can be viewed inthe appendix on page 97.]

Senator SANTORUM. Thank you, Mr. Secretary. I appreciate that. I am curious. You mentioned your tabletop exercise. First off, can you tell me, with respect to avian influenza and the outbreak, can you tell me some of the lessons learned from that exercise? And then I will see if you guys can tell me whether you are doing similar kinds of exercises and what you are learning.

Mr. WOLFF. I think the tabletop exercise really helped us understand that we need better personal protective gear. For example——

Senator SANTORUM. For whom?

Mr. WOLFF. That is exactly the point. We had PEMA at the table and we had the Department of Agriculture at the table. So if you have an outbreak of H5N1 in a large poultry operation, who goes in and depopulates the birds? Do you ask PEMA to go in and depopulate the birds or do you ask the Department of Agriculture to go in, and who has the equipment and the expertise to do it? Well, as it turns out, PEMA has the equipment and the Department of Agriculture has the expertise. Therefore, we did not have people trained to be able to go in and depopulate a large poultry population if they had H5N1—lesson well-learned, something you think that would really jump right out at you, but until you do the tabletop exercise and start to delegate those responsibilities, it becomes obvious where your shortcomings are—better communications in working together, notifying the agencies, who has authority, what are the legal authorities? If you are going to depopulate a farm, who is going to sign off on that and say, "You have the authority to remove $1 million worth of birds from this farming operation"? So they are the kind of issues that become obvious, that we need to have a better plan and more details as we work through this process so that we are, in fact, prepared if we were to have an outbreak of H5N1.

Senator SANTORUM. OK, Chuck, who has the legal authority and why does PEMA have the equipment and USDA have the knowledge?

Mr. CONNER. Well, Mr. Chairman, I am not General Counsel for the Department of Agriculture, I am the Deputy Secretary, but I will tell you that when it comes to issues of eradication and control, the Department of Agriculture has a lot of experience in the State of Pennsylvania and a lot of other States, and we believe we have full legal authority to take the actions that are necessary to control and eradicate the disease. That is not a statement going forward. That is a statement I think we have demonstrated a number of times in the past as we have successfully eradicated low-pathogen avian flu in a number of circumstances, including the State of Pennsylvania, some high-pathogen H5N2 circumstances, as well, that could have been more challenging, were successfully controlled and eradicated, as well. So we stand, I think, pretty well on our legal authority to do this.

Having said that, I think the Secretary of Agriculture for your State makes a very good point in terms of the need for full cooperation. It is absolutely essential when you have a situation develop like this, because the U.S. Department of Agriculture are not the only people on the ground. Obviously, we collaborate greatly with local people, the local laboratories that are there to do the testing. These are the types of areas I think where we feel like we have made tremendous progress in the last 3 years, is that coordination effort which is so essential for rapid response to these circumstances.

If you have got a particular product in question that is being tested at a State laboratory, there is no time for the results to sit around and to wait, and perhaps that might have been possible before, but with the relationships we have developed with private sector, but particularly the State laboratories, implementing them into our laboratory network, those types of things I just do not believe were possible going forward. We will know quickly when there is a problem. We will be prepared to move quickly, and, of course, that is absolutely key to a successful eradication effort, is getting it quickly.

As I noted in my oral testimony, the agriculture and food system we live in has food and product moving very, very rapidly, and that is to our benefit, but at the same time that is our challenge in terms of eradication because we must get hold of this product-to-product question before it gets fully into commerce.

Senator SANTORUM. The Secretary has talked about a simulation that he did. Are you folks doing simulations? Is maybe HHS, in particular, doing simulations with respect to avian flu, or are you doing it on the bird side? I know you might be doing it on the human side. Now that this outbreak is potentially present with us, what are you doing with it?

Mr. CONNER. At the Department of Agriculture—let me just say the dates. We can get you the precise dates and the information, but we had a major simulation exercise. I believe the title of it was Crimson Sky, I believe, about 2 years ago, and in that particular circumstance foot and mouth disease was the problem. The exercise was highly successful. You mentioned specifically avian flu. As you know, Mr. Chairman, the White House held an important simulation exercise to review our multi-agency and White House response in an avian flu outbreak. The Department of Agriculture played a very, very active role in that, as did the Department of Health and Human Services, and we are continuing a lessons-learned process in that exercise, as well. The Department of Agriculture, as I know other agencies do, do have additional exercise-type activities planned going forward so that we absolutely——

Senator SANTORUM. From the lessons learned, can you talk about any particular holes and anything that we in Congress may be able to help you with?

Mr. CONNER. I think from our standpoint communication and coordination are always the first step in that process. As Secretary Azar pointed out, just between our two agencies there are a number of areas of shared jurisdiction. Some of the jurisdictional lines, you sit back and you wonder how did that jurisdictional line end up where he has feed, we have meat, we have poultry, he has food products—those kinds of situations. Regardless of where those lines are drawn, we need to work together, and whether it is coordination activities, sharing information, although I think the lesson learned—what you learn in these exercises—is that need. Going beyond that, though, again, as I pointed out earlier, I think the progress that we need to continue to focus on is not just among our agencies, but reaching down to the State and, in this case, even the private sector for those vulnerability assessments so that we are not just sitting back, waiting on the event, but we are identifying where they are likely to occur and how they are likely to occur and take preventative mitigation measures hopefully before that happens.

Mr. AZAR. That is what I think one of the critical lessons learned coming out of the tabletop that the White House organized for the Cabinet members before the holidays on avian flu—pointed out the importance that, when it comes to preparing for a pandemic avian influenza outbreak in the United States, this cannot be the Federal Government alone. This is too big. If we look at just comparing an avian influenza pandemic in the United States to something as devastating as Hurricane Katrina, and compare the differences between them, Katrina, as horrible as it was, was confined in region and duration. A pandemic influenza outbreak would be nationwide by definition, with outbreaks occurring in multiple cities at the same time, and it would not last a couple of days. It would last for months, possibly 18 months. We do not quite know.

This is something that would bring every part of the American economy to bear, the Federal Government, State, local, private sector, all coming together to work on it. And that is why, coming out of this tabletop, the President directed Secretary Leavitt to convene pandemic influenza preparedness summits in every single State, and so right now HHS is working with USDA and the Department of Homeland Security and the Department of Interior. We are convening with Governors in every State pandemic summits where we go through what should State plans include, what types of industry ought to be at the table, how can everybody prepare for an avian outbreak, and importantly our request that every single State exercise those plans. A plan on a shelf is not useful. It has to be exercised. It is only when you actually run through it that you can identify the holes in it and the flow of information, learning, for instance, the importance of public communications and the accuracy and timeliness of communications to the public during any type of public health emergency.

Senator SANTORUM. Since we are on the topic of avian influenza, I know there are news reports today about deaths in Turkey. Do you have any update on that?

Mr. CONNER. Senator, I do not, and I do not know whether Alex does, as well. I read the same news clips as you probably did this morning with regard to the family, the one family, in particular, that had direct contact with the handling of the birds, and I have not gotten an additional update since early this morning.

Mr. AZAR. The latest information that I have—and it was going to be updated actually this afternoon with some more recent information from the World Health Organization—is that so far the cases that have been identified, as Secretary Conner mentioned, appear to have been ones where transmission would have been as a result of close, repeated, proximate handling of poultry, but we still need to continue to look into this. So at the moment it does not appear that the cases that have been identified so far are the result of efficient human-to-human spread of the avian influenza virus, but we do still need to work with the Turkish authorities and the World Health Organization to get all of the facts on the ground there before we state anything conclusively.

Senator SANTORUM. Mr. Secretary?

Mr. WOLFF. Well, I think that is kind of an interesting point, that as H5N1 starts to be spread through migratory birds, they feel that is how our greatest vulnerability of it infecting the poultry population in the United States could occur. It would basically go up through Asia and then come through Alaska, and then come down through Canada and infect the poultry flocks in the United States. So that is really why we feel that it is just so critical to have a good surveillance program in place, as well as a good response plan that, unfortunately here in Pennsylvania, has not been on the shelf, because we have exercised it a number of times. Fortunately, it has not been H5N1. It has been a low-pathogen—actually a non-reportable type of avian flu in every case. So we are very pleased that we have not had to deal with an H5N1, but the real threat for a pandemic in the United States obviously is that if it comes into the migratory birds, into the poultry population, then it can actually mutate to the point where it can be human-to-human population—is one of the ways that a pandemic could happen in the United States. But more likely the way it would happen would be that mutation would take place in another country and someone would travel to the United States, and then it would start to spread that way. But having said that, it really does, I think, come to the point as to how important it is to have funds in place to make sure that there are good surveillance programs out there for the domestic poultry population and have good response plans in order to be able to contain that if we were to have the H5N1 in our poultry industry here in the United States.

Senator SANTORUM. Are you satisfied that there are funds available for such surveillance?

Mr. CONNER. We are, Chairman Santorum, and are so because of your good work and the good work of the Congress in recently passing the supplemental request that Secretary Azar mentioned. The President requested those dollars from the Department of Agriculture standpoint. We actually requested and received the full $91 million that we sought. A good portion of that money is for surveillance, including surveillance of the wild bird populations coming down through Canada. And so we are confident that we have the mechanisms in place, particularly with this additional money, to achieve what I think are high-priority surveillance issues at this point.

Senator SANTORUM. If a migratory bird were to be tested and found, would that be something that the public would be made aware of or is that something that——

Mr. CONNER. Yes, we would, as soon as confirmation of that test would happen. Certainly that would be public information. I will tell you Secretary Johanns believes firmly in an open system of communication with the public. We have been through a number of issues, as you know, with BSE, the so-called mad cow disease, in terms of release of public information of that. I think we have demonstrated that we are not going to put out bad information prematurely, but as soon as information is confirmed we are going to make that information known to the public so that they can take appropriate action, whatever that may be.

Mr. AZAR. And Secretary Leavitt and the President—in terms of human-avian influenza preparedness, transparency and openness are the centerpiece of how America is trying to deal with this problem, and encouraging other Nations around the world to similarly be transparent and open. So obviously we have to set the example.

Senator SANTORUM. Very good. Thank you, gentlemen. I appreciate your testimony.

If the second panel could come forward? Thank you, gentlemen. I appreciate you being here. Let me go in order of testimony. First we have Dr. Robert Steele, who is the Dean of the College of Agriculture Sciences at my alma mater, Penn State University, and an appropriate blue and white tie that you have on, Dean Steele. Thank you for being here and for certainly the great collaborative effort that you have shown to me in our office and the work that we have done on agriculture policy and agriculture research, you and the school have been terrific in providing lots of resources to us and my staff, and I just want to thank you publicly for the cooperation.

Dr. Gary Smith, who is a Professor of Population Biology and Epidemiology at the University of Pennsylvania School of Veterinary Medicine. Thank you very much for being here.

Finally, Dr. Bernard Goldstein, who is the Professor of the Graduate School of Public Health at the University of Pittsburgh. So we have Pittsburgh, State College and almost Philadelphia represented here today. So thank you, thank all three of you, for being here.

Dr. Steele?

STATEMENT OF ROBERT STEELE, DEAN, COLLEGE OF AGRICULTURAL SCIENCES AT PENN STATE UNIVERSITY, UNIVERSITY PARK, PENNSYLVANIA

Mr. STEELE. Thank you, Chairman Santorum, and I appreciate the opportunity to represent the Commonwealth's land-grant university at the hearings here today. As you know so well, we have been celebrating our 150 year sesquicentennial, along with Michigan State University, as the Nation's two oldest colleges of agriculture, Penn State having awarded the first undergraduate and graduate degrees in agriculture in all of the United States, and we are quite proud of that 150–year history. And throughout those 150 years we have been dealing with biosecurity. We defined it different ways, depending on the decade.

We also are proud of the fact that of the nearly 150 4–year institutions of higher education that we have in the Commonwealth, that Penn State, as the land-grant university, is the only university that offers programming across the entire food and fiber spectrum, reaching into each of the 67 counties of the Commonwealth.

In my written text, I offered five examples or sample areas where we are working in collaboration with USDA, Pennsylvania Department of Agriculture, Pitt, as well as Penn, to illustrate a few key points, and those areas were avian influenza—and we have already been talking about that—the development of molecular biosensors, where we are taking the fascinating technology of biotechnology and merging it with nanotechnology, the area of plum pox, which Secretary Wolff mentioned, the issue of soybean rust—very, very important and key across the entire country—and then finally what we called the EDEN program, and I notice that it is in Secretary Conner's written remarks, the Extension Disaster Education Network that plugs into each of the 67 counties of Pennsylvania, along with PEMA and FEMA, but also into the nearly over three thousand counties across the United States.

Across those five example areas, I think a few common things emerge. First of all, we do not have all the answers for the issues of today, let alone what tomorrow's issues are. We need to continue to find ways to close these knowledge gaps that are there, and we need to close these knowledge gaps, whether it is the threat of an intentional introduction of a pest or pathogen or the unintentional or natural introduction of a pest or pathogen, because as a scientist I can tell you the science is the same. The science is the same whether it is intentional or unintentional with any pest or pathogen.

Second, and I believe probably the most important issue here when we talk about security, agriculture, biological security, is that this is a generational issue. This is going to be with us for decades, just as it has been. We will define it in different ways in the years ahead, but we have a very real workforce issue. Think back to the anthrax issue of a few years ago and how difficult it was to find people who knew much about anthrax. I know people at the FBI and the CIA came to Penn State and we went to retired faculty to bring them out to deal with this issue.

I saw figures just yesterday where 8 years ahead we are projecting the need for 24 million college graduates in the job market, and the demographics are predicting we are going to have 19 million. So we have this general workforce issue to start with in the area of agriculture security, biosecurity. These best and brightest young people, what can we do to convince them that they can have a rewarding career in this area, an important career, because yes, it is in the public good. As we all know, unfortunately, it is also an issue of national defense, as well.

When I look within Pennsylvania, I see the single-most important issue facility-wise, and Secretary Wolff already pointed this out, in biocontainment and the need for additional biosafety Level–3 facilities, not only at Penn State, but at Pitt and Penn and frankly wherever we need them. The one that will be coming into Harrisburg is just simply not going to be adequate, and we are woefully behind other States in the United States in this, and we need to pay attention to that.

At the national level, in my role——

Senator SANTORUM. I hate to interrupt your testimony, but could you explain while you are on the subject what the difference is between a Level 3 and a Level 2, and why it is important and why we are behind?

Mr. STEELE. Well, the different levels are a function of the nature of the pathogen, all the way up to a Biosafety Level–4—Ebola virus, for example, into a Biosafety Level 4. So it is an issue of the degree of containment and the measures and steps that need to be taken to protect you from the pathogen, the pathogen from you, and as you might guess, the higher the number, the price tag goes up. That is the short answer to your question.

Nationally, I think the issue we have—and we have been working with Secretary Conner on this, is we all know we are resource-constrained in everything that we do, and we need to find more effective and efficient ways of using the resources that we already have, and in my role—in a leadership role in the national land-grant community, there has been a small group of us working with USDA and others to fundamentally restructure our research, education and extension arm of the USDA to make better and more efficient and more effective use of the resources that we already have. I do think that there are some efficiencies that can be gained.

So in the end—and I know your subcommittee actually will become very important with us because this is Title VIII of the farm bill, which your Research Subcommittee deals with front and center. So finally I close with trying to ask the question, as this session is closing, are we prepared, and my answer is not nearly well enough because of the knowledge gaps, the workforce needs as we project in the future. And we need to get to work in doing this because if not us, who is going to do it? And so again, Senator, thank you for giving me the opportunity to present my testimony here with you today.

[The prepared statement of Mr. Steele can be viewed in the appendix on page 102.]

Senator SANTORUM. Thank you, Dean Steele.

Dr. Smith?

STATEMENT OF DR. GARY SMITH, PROFESSOR OF POPULATION BIOLOGY AND EPIDEMIOLOGY, UNIVERSITY OF PENNSYLVANIA SCHOOL OF VETERINARY MEDICINE, KENNETT SQUARE, PENNSYLVANIA

Dr. SMITH. Good afternoon, Mr. Chairman. My name is Gary Smith, and as you know I work at the University of Pennsylvania. Penn has a number of institutions and the initiatives that deal with all hazard response, but I am just going to concentrate on the particular responsibilities of the School of Veterinary Medicine, which is where I work. The school's primary contribution to the national effort to maintain agricultural biosecurity involves first training the veterinarians who will be among the primary responders to any threat to agricultural biosecurity, and then the research and service components of our work that is exclusively directed toward biosecurity issues. And that service component includes the National Salmonella Reference Center, the William Boucher Field Service, the Poultry Diagnostic Laboratory, and our contributions have already been mentioned to the Pennsylvania Diagnostic Laboratory System, and that is indeed the bread and butter of what we do, and it represents an existing infrastructure upon which our ability to coordinate research and to develop an integrated strategy to protect the agricultural industry from both accidental and deliberate introduction of pathogens in part depends, and that infrastructure, of course, is being elaborated all the time.

One example is the Emerging Primary Animal Healthcare Specialist Training Program, a long title for something quite simple. It's designed to train those who actually work on, for example, dairy farms, to recognize diseases and conditions that may affect productivity and perhaps some serious breakdown of biosecurity. But is agriculture prepared to defend itself? Well, up to a point. I am going to use just one example to illustrate what I mean by that. It is certainly the case, for example, that research workers like myself and my colleagues at Penn State have been extremely active over the last three or 4 years in exploring things like the mathematical models of foreign animal diseases that have proved so useful in controlling, say, mad cow disease and foot and mouth disease in Europe, and which could equally well be applied here in the United States.

It is absolutely no exaggeration to say that Pennsylvania, New Jersey and Maryland all taken together have more research groups with more experience and more expertise in areas like infectious disease modeling and syndromic surveillance methodologies than anywhere else in the United States, but the data that drive these systems are just not there in any kind of useful quantity.

Multi-infectious disease models absolutely depend upon something quite simple, knowing where the farms are and how the animals move between them, and that means we need good maps and we need an effective national animal identification system, and at the moment, with some caveats, we have neither. There are, indeed, current attempts to remedy these deficits at both the State and Federal level, I know, as I am sure you know, too, but I honestly am pessimistic that any of them will return useful information in a timely manner. And to be quite honest, our current research strategies—that is to say my own and my colleagues at places like Penn State—is to assume that we will never have these data and to decide to focus instead on workarounds, to attempt to use what is available in the public domain and to see if we can answer worthwhile questions about prevention and control using information that is much more granular, to use the jargon, than we would like.

Just in case you think this is some arcane issue of interest only to those who model infectious disease dynamics, it is worth pointing out that the perfectly ordinary methods of dealing with infectious disease outbreaks also depend on being able to trace animals back to their original farm. At the moment, this is done by questionnaire and by following the commercial paper trail. It is tedious, it is often inaccurate, and worst of all it is slow. It is very, very slow, and as several of the witnesses have already stated, speed is everything in these matters. The delay of just 24 hours can mean an epidemic of almost twice the magnitude.

One of the greatest impediments to the acquisition of good maps and of a national animal identification system are the quite proper and very realistic objections of the industry itself. There are some very complicated issues of confidentiality here, not to mention the required reassurances with respect to possible litigation that can only be resolved, I think, by legislation.

Thank you. I will take questions at the appropriate time.

[The prepared statement of Dr. Smith can be viewed in the appendix on page 108.]

Senator SANTORUM. Thank you very much, Dr. Smith.

Dr. Goldstein?

STATEMENT OF DR. BERNARD D. GOLDSTEIN, PROFESSOR, GRADUATE SCHOOL OF PUBLIC HEALTH, UNIVERSITY OF PITTSBURGH, PITTSBURGH, PENNSYLVANIA

Dr. GOLDSTEIN. Thank you, Chairman Santorum. I appreciate the opportunity to be here. I am a physician, until recently Dean of the Graduate School of Public Health at the University of Pittsburgh, and I will approach this testimony from a public health background. I ought to tell you that I'm also the former—under President Reagan—the former Assistant Administrator for Research and Development at EPA, which is a very long title.

[Laughter.]

Senator SANTORUM. We still think you are important here today, Dr. Goldstein.

Dr. GOLDSTEIN. Thank you. We in public health have been particularly concerned with food safety and with the public health infrastructure. Our Graduate School of Public Health is very research oriented. We are first in the Nation among State-related schools of public health in competitive NIH funding, only behind Johns Hopkins and Harvard among the 37 schools of public health.

Our research grants from NIH funding lead us to approach the terrorism issue.By addding onto existing biomedical capabilities. We have a lot of opportunities to do that. We have moved forward in a number of different ways in an attempt to do so. But we would note to you that the silos that separate out research funding streams in this country tend to make it difficult to move across the various different potential funding pathway. In my written testimony, which I am paraphrasing, I have given you an example of an opportunity based upon NIH funding to much more rapidly detect and much more cheaply detect new pathogens, that one of the Nation's experts has been unsuccessful in being able to generalize from NIH funding toward potential agricultural bioterrorism threats.

Our schools major interest, however, is in the rural public health infrastructure and trying to work to safeguard rural public health. We have worked along with our colleagues at Penn State and the University of Pittsburgh at Bradford in the rural areas of our State. We have developed a nationally known Center For Rural Public Health Practice. The response to bioterrorism has been central to this program.

We have had a major national meeting on the rural public health agenda, which we hosted through this center. It was interesting among these national experts terrorism toward the agricultural infrastructure was a major issue. Of the many different bullets that were in the document, the nural public health community noted, the importance that the locus of agricultural production could be key to bioterrorist targeting of the United States.

The programs we have developed at Pitt-Bradford have been reasonably successful. Our overall center in Pittsburg, funded by the Centers for Disease Control of the U.S. Public Health Service—our Center for Public Health Preparedness, was awarded over $5 million for the next 5 years to train front-line public health workers and first responders. We offer onsite and long-distance training, preparedness nursing, bioterrorism, preparedness law, forensic epidemiology and environmental health to a combined total of more than 700 public health professionals and clinicians. However, in collaboration with the Center for Rural Public Health Practice at Pitt Bradford, we could and should do more that is specifically aimed at rural public health practice and agricultural issues amidst this broad range of efforts related to preparedness, in general.

Our task is quite challenging for two reasons; one is obviously the lack of funding, but another has to do with the lack of public health infrastructure in our Commonwealth's rural areas. The Federal HRSA figures show that Pennsylvania ranks 50th among all States in our public health infrastructure. In the size of our public health workforce per capita. are number 50. It is particularly notable in our rural areas. As just one example, many of our northern tier counties have an infrastructure in place of perhaps a total of two public health nurses, while demographically similar southern tier counties from New York State have a public health workforce that counts in the dozens.

We need to do more to beef up the public health infrastructure in our State in rural areas to be able to respond to these threats, but our academic program has not waited for the Commonwealth to improve this infrastructure. We have used our CDC funding to reach out to the entire responder community—that is, firefighters, police, EMT personnel and others active throughout the rural parts of the State, and we have done this in collaboration with our colleagues at Penn State. Some of these programs are beginning to be copied nationally.

Finally, I should point out that there is a need for integrated assessment and planning. You have asked some very good questions about that earlier. There is a Center for Biosecurity at the University of Pittsburgh Medical Center. It is certainly the first of the major centers working on the issue of bioterrorism from infectious disease, but it is aimed at understanding the broad issues related to the threat of bioterrorism to the Nation's health, economic and social and political security. This began at Johns Hopkins. It has moved to the University of Pittsburgh Medical Center. Dr. D.A. Henderson is the physician who has been most responsible for the eradication of smallpox. He is the former Dean of the Johns Hopkins School of Public Health, and who was President Bush's key person in the response to the anthrax episode. We believe that there is really a need for better integrated assessment that should come to a large extent from reasonably independent sources. We think the universities of our Nation can provide not only that independent resource to look across these various different areas, but to help with generalizing to the research that is needed to anticipate the future threats that will come, as Dr. Steele says, for the next generation.

Thank you, and I welcome the opportunity to answer questions.

[The prepared statement of Dr. Goldstein can be can be viewed in the appendix on page 114.]

Senator SANTORUM. Thank you, Dr. Goldstein. Well, let me just pick up on where you left off, which is coordination. I mean, we have heard Secretary Wolff and the other secretaries talk about the one area that seems to come to the top, is communications and coordination. How much coordination and communications do the

USDA and Pennsylvania Department of Agriculture have with your organizations, and how involved are you in that?

Mr. STEELE. At Penn State, in the Pennsylvania Department of Agriculture, I think some of my colleagues sitting behind me would say on a daily basis. Our leadership teams meet on a regular basis three or four times a year on building a common agenda. Our scientists again are interacting daily. We are quite appreciative of the highway money that has gone in to make the road between State College and Harrisburg a little quicker to get through.

Senator SANTORUM. Almost done.

Mr. STEELE. Almost.

Senator SANTORUM. Almost done.

Mr. STEELE. But it is just an extremely, extremely effective cooperation which then extends largely through the system to Penn, and then on other issues of commonality to institutions like Pitt.

In a more separate, but parallel vein, through the land grant system where we are organized regionally, the Northeast region, the North-Central region, the West and the South, we interact with USDA primarily through what we call CSREES, the Cooperative State Research Education Extension System, and that is a very, very formal alignment, and it is, in fact, the essence of the land grant system.

Dr. GOLDSTEIN. Well, it is a great advantage to the University of Pittsburgh to be able to work closely and cooperatively with Penn State on a lot of these different issues. I have focused on our areas that we would like to be able to do more, and they range, as I say, from basic research to my own personal interest in research, which is on some of the issues that are basically abuse of public health principles in world trade to the detriment of American farmers. Again, these are the kind of things in which the collaboration could be enhanced, but it gets out of the bioterrorism area.

Senator SANTORUM. Well, you both mentioned the issue of worker training and the importance of that in our response, and I assume also in our notification, detection and the like. You mentioned the public health issue, our public health workforce. Just a question as to why you believe that we are in such—compared to New York and I assume other States, since we are 50th—that we are in such a bad state here in Pennsylvania. I mean, that somewhat surprised me, given other States that we are comparing ourselves with. So I would ask the question why, and then to Dr. Steele just a general question on the issue of workforce training and things that we might be focusing on here. My guess is we will be doing a major piece of education bill following up on—there was a commission that was put together headed by the former head of Boeing, Norm Augustine, who put forth ideas on how to get more kids interested in math and science and the importance of that. So my guess is we will be doing something along those lines. So if you can think about ideas that might respond to this issue in that context—can you talk about the public health issue first?

Dr. GOLDSTEIN. I'm in the Commonwealth 5 years now. I was surprised at those figures, as well. We have spent a lot of time and effort in our Center for Public Health Practice trying to understand why. Part of it is now you count the numbers. The State Department of Health would tell you, "Well, it is not quite as bad as that looks," but now you do the count does not explain it all. We are so far behind everybody that even if you counted the public health workforce somewhat differently, we would still be 48th or 47th.

The key issue from the point of view of this hearing is that the problem is mainly in our rural areas. We have some very good public health workforce groups in Philadelphia, Allentown, and other areas—Pittsburgh, certainly, Allegheny County. The professionalism of the State Department of Health is excellent, in my opinion, but there are just too few folks and they are too busy, and when you get anything added to it, they are simply not going to be able to respond.

Senator SANTORUM. Well, I guess my question is why is the funding in Pennsylvania so disparate from the funding in New York? I mean, is it State policy?

Dr. GOLDSTEIN. It is our States that make the decision.

Senator SANTORUM. OK.

Dr. Steele?

Mr. STEELE. Senator, the workforce issue, I think—and I am going to confine my comments to the college graduate—obviously, it is much, much broader than that, but a young person graduating from college has options, as I pointed out, just simply looking at the numbers, and fundamentally they are going to go where they see opportunity, and our challenge is at the youngest age possible, and the literature would suggest by the third or fourth grade if you want to get a young girl interested in a career in math or science. So we need to start as early as possible to help these young people understand that there are attractive, viable careers out there in areas such as biocontrol, security, molecular biology, nanotechnology, whatever.

As I said in my comments, it is in the public good to do that, and particularly in the food and fiber sector. And you can never do enough of it, and the numbers and the demographics are, quite frankly, working against us right now.

Senator SANTORUM. Yes. One of the things that has struck me in the these reports is showing that the number of engineers that we are graduating has basically been flat over the past couple of decades, which China now graduates almost 10 times more engineers than we do. If you consider the fact that technology jobs are the future of any economy, what economy is going to succeed when China, India and others—and even countries much smaller than we are, are graduating more folks in the technical fields.

That is a big issue for us. If you have any comments on specifically how we deal with that, I would be anxious to hear them.

Mr. STEELE. I, a few months ago, was in China and had the opportunity to address a freshman class at a university there, an agricultural university of 9,200 young people, 9,200, and we read in the papers about the hot China economy growing at eight, 9 percent clips. When you go over there, you see it right in front of your eyes, the investment that is being made in their educational system, and their goal, which was so clearly stated to me, was to have 50 percent of their 18–year-olds go on and get a university education versus—and I believe it was something on the order of 18 percent right now, and they are just investing heavily into their science and engineering infrastructure. And these young people are understanding and seeing that there are career opportunities available, and that is in part why we are not seeing as many of them coming to the United States, as well.

Senator SANTORUM. Very good. Thank you, gentlemen. I appreciate your testimony.

While the final panel is coming forward, let me do a couple of housekeeping things. First, I had asked the President of the Pennsylvania Farm Bureau, Carl Shaffer, to be here today to testify, but he is out of the State on another mission and could not be here. I just wanted to thank him for taking the time to prepare a statement, and I would like to enter that statement in the record and thank him for the great—again—cooperative work that we have done with him and the Pennsylvania Farm Bureau in representing the interests of Pennsylvania agriculture.

[The prepared statement of Mr. Shaffer can be viewed in the appendix on page 134.]

Senator SANTORUM. Second, I just want to introduce State Senator Noah Wenger, who is here today. Senator Wenger is a member of the Senate Republican leadership in the State Senate and is a great friend of Pennsylvania agriculture, former chairman of the agriculture committee in the Senate of Pennsylvania. So thank you, Senator, for being here.

Our final panel introduced—first is James Adams, who is the President of Wenger Feeds. Thank you, Mr. Adams, for being here; Walt Peechatka, who is the Executive Vice president of PennAg Industries; Lew Gardner, who is a dairy producer from Galeton, up in the great northern tier; and Keith Masser, who is the President of the Pennsylvania Cooperative Potato Growers. Thank you all for being here.

Mr. Adams, would you proceed?

STATEMENT OF JAMES ADAMS, PRESIDENT, WENGER FEEDS, RHEEMS, PENNSYLVANIA

Mr. ADAMS. Thank you, Chairman Santorum, and thank you for putting this as a high priority on your agenda and visiting the State at our State Farm Show. Wenger's is a 60–year-old regional feed manufacturer that mostly delivers to poultry and swine farms, and we act as sort of a conduit for local grains and soybeans, convert those into a high-quality feed, and then distribute those out to numerous family farms. Sometimes we have up to 500 locations that we deliver to, so biosecurity has been important for us for many years, and specifically, having gone through the 1983–1984 avian influenza, we have disinfectors on our trucks and our servicemen practice very high standards of biosecurity. So we are very familiar with the on-the-farm, on-the-ground type of biosecurity that is important in this.

The reason we have that is we need to protect our customers, whether they are a large corporation or they are a family farm, that we are not tracking around any type of disease. Also we own millions of laying hens ourselves, we want to ensure that we do not lose an income from that. So it is a very important subject to us.

As I mentioned, we learned a lot of lessons in the 1980's from avian influenza. I guess I am a little more optimistic than some of the other panelists have been, in that I have seen what farmers and agricultural people can do in the face of an emergency. The industry in Pennsylvania actually was credited with jumpstarting the effort to control avian influenza in 1983 by getting together cooperatively, even though they were competitors, and getting a lot of information together, where farms were, although at that time we did not have GPS systems. We actually did it with cardboard and county maps and sticking little pins in and things like that, but the industry did urge our local university and the PDA to put together a GPS system that we have used since 1983–1984. It actually gets instituted rather rapidly when an outbreak occurs and the industry again works cooperatively to try and address these situations.

Anything can be improved, and I guess the message for you, if there is help that we can have, is basically in four areas, and they would be: detection, indemnity, vaccination and protection. With detection, we do have a relatively good system, and I think Secretary Wolff's written comments showed that we have about 240,000 AI samples that are submitted every year, that the industry is trying to safeguard and have a knowledge base of what is going on here.

Those tests can always be upgraded. They need to be more rapid. They need to be more accurate. A lot of them tell you that you have an AI, but they do not specifically tell you what kind you have. So we can use better tests at any time. They should be very quick and hopefully inexpensive, so that they can be put into action rather quickly. When you have flock sizes of 200,000 or herds of 2,500, you need to have economical tests, also. With the indemnity program, I think Pennsylvania farmers are not coming with their hand out and saying, "Bail us out of this," but what a good indemnity program does is it promotes people cooperating and joining a surveillance program, rather than just taking their animals to the normal market, driving down the road and spreading disease along the way. So it needs to be a specific program, but it needs to be effective enough that it gets the cooperation of most of the people that are in the industry, mostly to encourage their submission of samples.

Another idea that goes along with that, that I think leverages the Government's money, is if we could have a subsidized business interruption program similar to crop insurance, so the Government is not paying the full amount, but they are paying enough that it encourages people to do that, because a lot of times, if it is one or two or three flocks, a company that is involved with that may be able to ward off that type of economic impact, but if three of those people are individual farmers and that is their sole income, they could be wiped out if a good indemnity program or a business interruption insurance is not in place.

The third area is vaccination, and Pennsylvania has been in the past prevented from using vaccination effectively. We did stockpile some. We allowed another State to use it. The stockpile for Pennsylvania has not been re-created, but we would encourage a seed stock and research into how you can quickly multiply those viruses into effective vaccines, because it will take purification and replication to get a vaccine that is in a usable form that will not harm the animals that are going to be using it, so it would be a very specific stockpiling of the vaccine.

The final area would be protection of the farms. I am not sure if the news got to you down in Washington, but we had a local animal-rights group break into a farm in the area, and that is trespassing, that is breaking and entering. And to be fair, local police are not sure what their boundaries are. State Police might not be familiar with agriculture. The local prosecutors may not know everything that they can bring to bear, and they may think, "Well, all they did was they broke into a chicken house, what is the big deal?" But if they were carrying disease into that farm and that became a reservoir for neighboring farms, or if they deliberately brought something into that farm, it could create a big problem. So we would like to see stiffer penalties on that, not have it be a misdemeanor. If somebody is actually breaking and entering and trespassing onto someone's property, whether they are a farmer or not, we think that should be seriously dealt with, and it will help the biosecurity efforts, not just the property rights.

We again appreciate you coming here to Pennsylvania and holding these hearings, and just need a little bit of fine-tuning in a lot of the programs that we have to leverage what we are already doing.

[The prepared statement of Mr. Adams can be viewed in the appendix on page 117.]

Senator SANTORUM. Thank you, Mr. Adams.

Mr. Peechatka?

STATEMENT OF WALT PEECHATKA, EXECUTIVE VICE PRESIDENT, PennAg INDUSTRIES, HARRISBURG, PENNSYLVANIA

Mr. PEECHATKA. Thank you. Good afternoon, Mr. Chairman. PennAg Industries Association appreciates the opportunity to offer comments this afternoon on behalf of our association. For your information, we represent more than 600 agricultural businesses, including, I might add, Wenger Feeds, as a member of our organization.

I would like to start by reiterating and reemphasizing some of the comments that Mr. Adams just made to you on behalf of the poultry industry because it is very important information. We believe Pennsylvania is the leader nationally in the area of biosecurity on our poultry farms. After a devastating avian influenza outbreak in early 1980's, Pennsylvania's poultry industry began an aggressive effort to prevent future outbreaks and minimize their impact if they did occur. All commercial poultry operations in the State have been included in a data base that can be utilized by the industry and by our regulatory partners in the USDA and the Pennsylvania Department of Agriculture in the event of a disease outbreak.

In addition to the excellent data base that we have available, the second component of the Pennsylvania program is the surveillance practice by our poultry industry. Each year, poultry producers send nearly 250,000 routine samples to our laboratories for diagnosis.

I would like to turn to a few recommendations, and we would encourage your subcommittee and the U.S. Congress to consider the following items in addressing the issue of biosecurity and developing other initiatives which provide financial support to the agricultural community. First, increase the diagnostic laboratory capability in States with large concentrations of poultry and livestock. We encourage the Federal Government to provide additional financial support for the development of a BSL–3 laboratory at one of our Pennsylvania research institutions, so that the facility can be used for both research and for routine sampling and diagnosis.

Second, we believe every State should be encouraged to develop a premise identification program along the lines of the program Pennsylvania has developed for the poultry industry. However, the issue of information confidentiality must be addressed when undertaking this effort.

Third, agri-terrorism has been occurring with increasing frequency. Animal-rights groups have been breaking into research facilities and commercial poultry operations for the purpose of releasing animals and filming the contents of these facilities. These activities are strict violations of the basic tenets of biosecurity and place the owners of these operations in jeopardy of having disease introduced to their operations.

To address this problem, we recommend that the Federal statutes be strengthened. Federal laws should be such that violations are recognized as a national threat to our country and to those that produce the food stuff to feed our population. I would also like to take this opportunity to commend the FBI for its ongoing efforts in the terrorism incidents in Pennsylvania. They have done a wonderful job with the existing statutes, and we believe that they could even do a better job if those statutes were strengthened.

Fourth, we recommend that a significant portion of any Federal appropriation for addressing the bird flu threat be devoted to fieldwork and adequately preparing the agricultural community to prevent a bird flu outbreak or to promptly contain it, should it ever occur. While I am on the subject of bird flu, I want to take the time to emphasize the fact that it is important to recognize why bird flu incidences have occurred in Asia and Eastern Europe. Most of the countries which have reported outbreaks to date have humans and birds and animals cohabiting in the same house or in very close proximity. These birds and animals run loose, and if they are carrying a virus they are spreading it rapidly throughout the house of its owners and perhaps throughout the neighborhood.

One of the many advantages of modern confinement animal and poultry operations, which we do have in this country, is that animals are confined in environmentally controlled barns, thereby minimizing the risk for infection. Confinement also eliminates the potential for wild birds and animals to spread the disease from flock to flock since the domesticated animals are isolated from the wild ones.

Fifth, one of the most frequent disruptions of poultry production in Pennsylvania is the problem associated with live bird markets. These markets in some of our major cities are reservoirs for disease, which can be easily spread back to the farms that are supplying the birds.

Last and No. 6, while most of our comments and recommendations today are centered on our animal and poultry agriculture, one remaining issue that also requires increased attention is our agrochemical industry. This industry, which PennAg also represents, provides the pesticides and the fertilizers to those in production agriculture. Agri-chemical facilities are also subject to the threat of terrorism. In this instance, also, we believe that the Federal statutes could be strengthened so as to discourage terrorism at these facilities. Penalties should be increased to minimize or eliminate the potential for terrorism against this industry.

We appreciate the opportunity to offer these comments, and I stand ready to answer questions you may have. Thank you.

[The prepared statement of Mr. Peechatka can be viewed in the appendix on page 122.]

Senator SANTORUM. Thank you.

Mr. Gardner?

STATEMENT OF LEW GARDNER, DAIRY PRODUCER, GALETON, PENNSYLVANIA

Mr. GARDNER. Thank you very much, Senator Santorum, for inviting me here and holding this hearing on biosecurity. I would like to take this opportunity as a dairy producer in Pennsylvania to thank you for your effort in MILC legislation that is in progress and hopefully signed by the end of this month. I am a dairy farmer from Potter County, Pennsylvania, and we operate a 250–cow farm with my wife and my two sons, and the safety and security of the Nation's milk supply is of the utmost importance to the U.S. industry and dairy producers in the United States, as well as Pennsylvania and all over.

The dairy industry partners have been working closely together for the last several years to develop initiatives that would address biosecurity challenges in the dairy industry. Biosecurity is an important aspect of ensuring a safe food supply and protecting our animals' health. My cooperative, Dairy Farmers of America, has taken a number of initiatives in recent years to secure their milk supply throughout the production and processing chain, and I have provided some details in my written testimony of that fact.

Dairy farmers have begun to adopt practices that will minimize opportunities to intentionally tamper with their livestock and milk for biosecurity measures. Common sense must be used when developing and implementing biosecurity practices on farms. In Pennsylvania, the vast majority, 85 to 90 percent of our dairy farmers, milk less than 100 cows per farm. Small farms will definitely need some kind of a financial resource to help them implement the biosecurity practices and install the technology and surveillance systems that is proposed.

About 25 percent of our farmers in Pennsylvania are, because of religious belief, Amish or Old Order Mennonites and do not use electricity. That should be taken into consideration. They are a vital part of our dairy industry. The implementation of Federal biosecurity regulations must be in sync with State regulations and other regulatory agencies. I will give you an example of that. As a dairy producer, my farm, my milk house and my milk tank must be open for inspection by U.S. Public Health 24 hours a day. If a U.S. Public Health rating officer comes to my farm and my milk house is locked and my bulk tank is locked, then I could receive a zero score on a U.S. Public Health rating which would jeopardize my ability to market my product. So what I am saying is common sense must prevail. I think U.S. Public Health says that if they have a key, then that is all right. There are 10,000 dairy farmers in this State and six rating officers. That is a lot of keys for a lot of people to carry.

Dairy product safety and security measures have been in place at processing facilities; PMO regulations, drug residue testing on every tankerload of milk, milk tanker seals and tanker wash tags. If any one of those seals is tampered with or absent, that milk is rejected. HACCP procedures are in place in plants to ensure appropriate plant operations, employee screening and background checks and restriction to access to milk plants. More work needs to be done on that. We need more work done on USAID programs. They need to be made mandatory and resources must be made available for the public sector to accomplish these program goals within 3 years.

Materials need to be developed to aid dairy farmers understanding the need for biosecurity practices. Penn State, through their extension program, could play an active role in this. National biosecurity practices must be developed and coordinated by individuals who understand the industry with the goal of implementing reasonable and practical guidelines. Federal and State resources must be made available to help dairy farmers offset the cost of implementing practices and installing technology that is available to accomplish this. Federal regulatory agencies should recognize the efforts and capabilities of their State partners in developing and implementing on-farm security measures.

In conclusion, the dairy industry partners recognize the need for enhancing biosecurity practices throughout the industry and look forward to working closely together with our Federal and State partners to accomplish our goals. I want to thank you very much for this opportunity to appear before you. I look forward to your questions or comments.

[The prepared statement of Mr. Gardner can be viewed in the appendix on page 125.]

Senator SANTORUM. Thank you, Mr. Gardner.

Mr. Masser?

STATEMENT OF KEITH MASSER, PRESIDENT, PENNSYLVANIA COOPERATIVE POTATO GROWERS, SACRAMENTO, PENNSYLVANIA

Mr. MASSER. Thank you, Chairman Santorum. On behalf of production agriculture, the Pennsylvania Potato Growers Co-op and the National Potato Council, I thank you for allowing me to testify amongst these esteemed guests. I am the eighth generation in my family to farm in western Schuylkill County, Pennsylvania, and if I assume—according to the last panel here, I am probably a detriment to our economy. I did acquire an engineering degree from Penn State, and after working in that profession for a couple years, I thought I could serve my family better by going back and joining the family farming operation. So sorry, you guys, but I did not apply the engineering degree to its fullest extent. However, we did try and use it. We built a processing plant, so I'm basically involved as operating three companies as president of our family farming operation, where we grow 3,600 acres of crops and we distribute potatoes throughout the Mid-Atlantic region.

Keystone Potato Products is a processing company that we just got launched this year through the help of some USDA initiatives and State initiatives. We are using methane gas to fire a boiler to produce steam to dehydrate potatoes. We are using these dehydrated potatoes and selling those ingredients to bakeries and pierogies and processed potato chips, and the Pennsylvania Potato Growers Co-Op is a marketing organization that we use to market growers' potatoes throughout Pennsylvania.

In our farming operation, we use a third-party food safety inspection program called the American Institute of Baking. We do it voluntarily to help support some of the Government agricultural practices that are initiated by the U.S. Federal Insecticide and Rodenticide Act. We market a lot of people's potatoes throughout the country in Pennsylvania, and we do require that they send us confirmation that they are following the GATT procedures, basically that they are using the pesticides according to label, that they have sanitary conditions on their farm and that their produce is safe. There is this program that we do voluntarily, then there are some customers, such as AlDI's Supermarkets and Albertson's—they put out an initiative saying they require us to have these third-party food safety programs in place, although they are not enforcing them. They are buying from suppliers who do not have these programs in place.

At Keystone Potato Products—we do have a third-party safety food program in place, and we chose a National Processors Association food safety program. We did this because of the many different third-party programs that are available. We chose the one that would encompass most of the companies we want to sell to. Campbell's Soups requires that particular auditing agency for us, so we selected it. Now, if we were to sell to Appleby's, they would choose Silica. These programs cost a lot of money, and we need to try to find within our industry some kind of consistency so that we can have one third-party agency that would encompass the different suppliers. That is a problem that I see surfacing through these third-party auditing programs, is that we could possibly have three third-party auditing programs at our company to service three different accounts.

How do we have the smaller producers that we are involved with have a third-party food safety auditing scenario? It is expensive. The primary expense is making it happen, but I have a full-time person on staff, and she is doing the recordkeeping, the inspection process, checking the rodent traps and making sure all the lights have safety protection, so there are glass covers so we do not have glass entering with product. It is tough—it is difficult for a small producer to be able to have a full-time person auditing their farming operation or their shipping operation. So one of the things that I can see coming about and that we would be initiating is have an organization like the Pennsylvania Potato Growers Co-Op select a third-party food safety auditing system and then help the growers that are members to perform these practices and have that take place.

What I think the Government—needs to take place here, is to protect from the accidental and forceful introduction of a pest. If one would remember back, the Irish Potato Famine was an event that significantly impacted our industry, and many of the Irish immigrants that came to this country were as a result of that famine, and that was caused by a fungal disease called Late Blight. We still have problems with that disease, and we need to continue with our research dollars to help keep that disease in check.

We need to have our borders protected from pests like potato wart and Ralstonia. Ralstonia is a disease that enters this country through—and it did recently—through geraniums, imported geraniums. This causes brown rot in potatoes and it could devastate our crop. I got concerned when I saw some personnel being taken from APHIS and getting taken and taken into the Homeland Security Agency. We need to make sure that people in the Homeland Security are trained to those standards that APHIS has in their knowledge to make sure we have our borders protected from the introductions of those accidental pests, and those types of pests could be introduced forcefully, as well. That could tear up our industry.

So basically those are the concerns I have, and I would able to answer questions at any point that is appropriate, and my written testimony is presented, and thank you very much.

[The prepared statement of Mr. Masser can be viewed in the appendix on page 131.]

Senator SANTORUM. Thank you, Keith, appreciate that, and thank you all of you for your much fuller written testimony. All of that testimony will be entered into the record. Just a couple of questions because we are running a little late—you mentioned, Mr. Adams, the issue of vaccinations and problems in your testimony. You mentioned problems with international restrictions on vaccinations. Can you elaborate as to what the problem is in that regard?

Mr. ADAMS. As I understand it, there are certain diseases that are recordable diseases, and we have been fortunate that they have split up avian influenza into just—the highly pathogenic ones are the reportable ones. But most countries do not want a country that they would import from to use an avian influenza vaccine, because what I have been told is they feel that they would not know when you have a real disease and when you do not, but we feel that with modern vaccines, one, you would use a killed vaccine which would not replicate; the other was, with some of the technologies, you can put a marker on the vaccine so that you could go back, and if you drew blood from that bird or got tissues, you would able to tell whether it was a natural infection or if it was a vaccinated virus that was in there. So we feel the technology has outpaced what the old trading laws were.

I know we have had conversation, and we had a presentation—I think here—for USDA to help us get that message across, and some countries just say, "Well, that is our law and that is how we are going to follow it." So I think it falls in the trading area.

Senator SANTORUM. Is it, in this case, just a matter of either a convenient way for a non-tariff barrier or a true misunderstanding of the science? Which do you believe it is?

[Laughter.]

Senator SANTORUM. OK. You have answered my question.

Mr. ADAMS. I think a lot of it is a trade barrier problem.

Senator SANTORUM. Yes. OK. You have answered my question.

Mr. ADAMS. Because some countries have used it very effectively. Italy used it very effectively, and there is no problem, so it has——

Senator SANTORUM. You say they used it?

Mr. ADAMS. They used a killed vaccine in Italy to stem their avian influenza, and also this last round up in Connecticut, the United States got to use vaccine, and part of that has been because they started to look—because of the size of the United States, they are looking more at regionalization, allowing that if you have avian influenza in one part of the country, it does not eliminate exports from the rest of the country. So we have made some progress in that area, but it is still—a lot of it is not based on science, and that is what we should go toward.

Senator SANTORUM. You mentioned the area of detection, and I know you, Mr. Masser, just talked about that, too. Have you noticed—and I know we just appropriated money for detection in avian flu—but have you noticed over the last few years since terrorism and bioterrorism has come much more to the fore, an improvement in detection of pathogens by the governmental entities in your field?

Mr. PEECHATKA. I agree with what Mr. Adams just said. Probably one of our researchers from the previous panel would be in the best position to answer that, but I think, by and large, there has been an increase.

Senator SANTORUM. I would like to know what the folks who are actually out there working feel, how they feel, whether you sense that there is a better system out there.

Mr. ADAMS. Well, as I mentioned, and I think several others had it in their written testimony, Pennsylvania is doing 240,000 avian influenza samples already. That is partly imposed on us by States where we want to sell our products to, but partly it is self-imposed, that the industry wants to know where avian influenza is. So for us to have greater detection, we would have to have more testing done.

I think you probably saw the headlines where the National Chicken Council just mentioned or just announced that they are going to be testing every flock that goes to slaughter for avian influenza. So that will be millions and millions of birds a year. People could criticize it and say it is after the fact, but AI can move so quickly that that is still an important piece of information that you can then work backwards in the regions and see what is going on. But to get more detection, I'm not sure. What we would like is that it is more rapid, because a 1–day delay in avian influenza management, once it is already here, and that management is so critical, and we learned that lesson too well. We started acting in Pennsylvania very, very quickly, and you could see from the data we went from—I think it was a $65 million-dollar price tag and 17 million birds lost, to the last outbreak that we had was one farm, 200,000 birds, and if it was $400,000, I would be surprised. So we have learned a lot of lessons. We are not perfect, and we could get surprised at any time, but we have a pretty good system in place, but you have to act very quickly.

Even in 1983–1984, and I know in some of the Delmarva incidents, the companies have taken the expense to put down the very first flocks, and they said, "We do not care about government help or anything. We just think it is the right thing to do," to act so quickly that they will just get rid of the first two sentinel flocks that are out there. Speed is what we need, so in detection we need a research into new detection methods that are very specific for H5N1 or H7N2 or H5N2, whatever the culprit is at that time, and not have a vague, "Yes, you have AI, but we really do not know what kind it is."

Mr. MASSER. I think the detection or the technology has gotten better, but logistics has gotten slower. For instance, we have a quarantine on golden nematode in New York State, which is a nematode pest that affects potatoes and potato production that was introduced from Europe back in the 1800's, and there is a USDA program that is in place. It is properly funded, but it takes a lot of priority to make sure it is properly funded, to quarantine and make sure that pest is in place.

So what I am talking about then is it requires samples. We get our farm sampled periodically to make sure that the nematode is not moving. So you test around the area to make sure that the nematode has not got introduced into neighboring counties and States. We just got involved in a government contract shipping potato flakes to Moldavia. It took 12 days for the Federal Government to test our samples and to get the results back before we could ship.

I do not think it takes that long to test the flake for microbes, but it is the logistics of making it happen that takes the time, and that takes money to have testing. Our samples had to get sent to Florida and have it tested and get the results back. So more testing facilities or having the technology being distributed so our products can get tested faster would be appropriate.

Senator SANTORUM. Mr. Gardner, you mentioned your concern about information on biosecurity—information materials on biosecurity. As you may have seen, Mr. Azar from HHS, representing FDA, held up a little pamphlet talking about biosecurity in the milk area. Have you seen that pamphlet before?

Mr. GARDNER. I had not read it, but I have seen it. We have pamphlets we hand out, a kit, to all of our producers on the different agencies, the different procedures that are being held, selling, restricted access and those kinds of things that we have done. That is the kind of information I am talking about, basic information for the producers would be important.

Senator SANTORUM. My question is do you feel like the Federal agencies involved, the universities involved, are providing enough information to you with respect to biosecurity in your operations?

Mr. GARDNER. I think the agencies are working together very well with providing the information, all the agencies together. I think the one handicap we have now is time. The animal identification process has been very slow in this country. Dairy—most of our animals are identified. It is just a matter of centralizing that information so that if there is an animal that has a disease, that it can be tracked rapidly, and all of the animals need to be tracked with the beef and all of the animals, but that has run a little slower.

It has come along, but we have to have a mandatory identification premise animal identification program that identifies the animal and the premises where it came from, and this takes coordination between the State and Federal Government and the other agencies.

Senator SANTORUM. Do you agree with that?

Mr. ADAMS. Yes, I think they have done a good job. It is a job that never ends. You get new producers that get involved, and older ones get out of the business. I think it is a three-pronged thing. We need governmental help maybe to produce publications and things like that. We need universities to help distribute that, because they have a lot of meetings already set up, and the last thing we need is another meeting. So they can tag a biosecurity issue into one of those meetings, but the third, I think, is the responsibility of the industry to either take the publications or make their own, or dissect them and translate them into things that producers actually understand, and a lot of it has to be a show-and-tell—you know, this is what you should actually do—because many times you will sit and, if a program is at the end of a long day of other sessions, you are going to miss some of that material. So I think it is a constant, never-ending educational process that we have to do.

Senator SANTORUM. Is the material being distributed in forms that are useful to you, not just pamphlets, but on the Internet, and that are accessible to you, maybe interactive?

Mr. ADAMS. Yes. Right now I would have to say—and I hate to always bias this toward avian influenza, but I will anyway—there are about maybe 10 sites right now that you could go to if you wanted to know what is happening in avian influenza, what biosecurity measures should you do. There is enough web sites out there that are very good. They have been put together, I would say, pretty rapidly, too, and they are very accurate, but you have to have the mentality that you want to go to the Internet to look for those things, and I would have to say that if the average age of a farmer is 58, I am not sure they are all going to be in touch with that, not that age is a determining factor in that, but maybe experience.

Senator SANTORUM. It is a good indicator.

Mr. ADAMS. Experience is, but I think there is a lot of material out there.

Senator SANTORUM. Animal identification?

Mr. ADAMS. In swine and dairy and beef, it seems to be a logical way to go. In the poultry area, we have a problem with thousands of animals in one load on a truck. We would look more toward facility identification, which there is some of that already, or maybe as a worst-case compromise is maybe some type of coop identification or something like individual bird identification. They have done some testing, and it then becomes a food safety problem, that the little tag or something that you put on gets in with the meat product because of the smaller size of a bird versus some of the other mammals that are out there.

Senator SANTORUM. Thank you, gentlemen, very, very much for your testimony, and we are adjourned.

[Whereupon, at 3:14 p.m., the subcommittee was adjourned.]

APPENDIX

January 9, 2006

**TESTIMONY OF THE HONORABLE CHARLES F. CONNER,
DEPUTY SECRETARY, UNITED STATES DEPARTMENT OF AGRICULTURE
BEFORE THE U.S. SENATE
COMMITTEE ON AGRICULTURE, NUTRITION, AND FORESTRY
SUBCOMMITTEE ON RESEARCH, NUTRITION, AND GENERAL LEGISLATION
JANUARY 9, 2006**

Good afternoon Mr. Chairman, I am pleased to be here to discuss Federal and State collaborative efforts with respect to food and agriculture bio-security. This is an issue of importance to our Nation, and one we take seriously at the United States Department of Agriculture (USDA).

As you are aware, I testified before the Senate Committee on Agriculture, Nutrition, and Forestry in July 2005 on food and agriculture security issues. My testimony today updates much of that information, but also includes more discussion of Federal-State collaborative research efforts.

USDA considers food and agriculture bio-security issues essential to its mission of providing leadership on food, agriculture, natural resources, and related issues based on sound public policy, sound science, and efficient management. The success of USDA's efforts is dependent on the coordinated work of a broad range of Federal, State, local, and private sector partners.

Food and Agriculture in the Context of Homeland Security

Agricultural exports were $62.4 billion in fiscal year (FY) 2005 and are expected to reach a record-high $64.5 billion in FY 2006 – with agricultural trade being a positive addition to our overall balance of trade. Agricultural imports are also significant with a value of $57.7 billion in FY 2005. Our nation's food and fiber system contributes approximately $1.24 trillion dollars, over 12 percent, to our gross domestic product and it employs about 17 percent of our entire workforce.

We face many challenges in protecting this important infrastructure. The agricultural industry is particularly concerned about security because diseases and pathogens—whether they occur naturally, are unintentionally introduced or are intentionally delivered—do not respect geo-political borders. The interconnected nature of the global food system contributes to our economic strength by improving production and marketing efficiency and providing timely responses to consumer needs. But, this integration is also a challenge in the event of attack or natural disease outbreak. The fact is, products move quickly from State to State and nation to nation; a pest, disease or other agent could spread just as quickly.

Since September 11, 2001, USDA has made great strides in expanding our mission to better include security. We remain steadfast in our belief that the threat to agriculture is very real. The Department has been working closely with its Federal, State, and local government partners, as well as with industry stakeholders, to address these concerns via a sector-wide strategy based on White House guidance.

We are relying upon guidance provided in Homeland Security Presidential Directive (HSPD)-7, *Critical Infrastructure Identification, Prioritization, and Protection*, and HSPD-9, *Defense of U.S. Agriculture and Food*, which address our preparedness for intentional acts of terrorism and the unintentional introduction of agents, pests, and diseases that could harm our sector.

HSPD-7: Critical Infrastructure Identification, Prioritization, and Protection

USDA has worked in coordination with the Department of Homeland Security (DHS) and our partners at the Department of Health and Human Services' (HHS) Food and Drug Administration (FDA) to ensure that we develop a coordinated approach toward implementing HSPD-7 for the food and agriculture sector. Central to this directive are the requirements for the government to collaborate with the private sector for infrastructure protection and to create an overarching framework and unique sector plans for protecting key assets and resources.

Since August 2003, USDA, DHS and HHS have worked with Federal, State, local, and private sector participants to establish a formal entity for sharing sensitive information, new policies, best practices and vulnerability assessments on a regular basis to help ensure the protection of the U.S. food and agriculture sector. The government entity, the Food and Agriculture Government Coordinating Council, is led jointly by DHS, USDA, and HHS, and includes Federal, State, and local officials. The private sector entity, the Food and Agriculture Sector Coordinating Council, is comprised of two leadership officials and an alternate from each of seven sub-councils representing the farm-to-table continuum. The two Councils meet quarterly and regularly hold individual and joint calls to discuss issues of mutual interest, such as sector vulnerability assessments and Federal research and development plans. As a result of the joint sessions, the Councils create working groups to address specific issues and report recommendations.

HSPD-9: Defense of United States Agriculture and Food.

In January 2004, the White House issued HSPD-9, which set forth the framework for a national strategy for preparing, protecting, and enhancing the security of the Nation's agriculture and food infrastructure. HSPD-9 states that a national policy must include programs addressing:

- Awareness and Warning;
- Vulnerability Assessments;
- Mitigation Strategies;
- Response Planning and Recovery;
- Outreach and Professional Development;
- Research and Development; and,
- Coordinated Budgets.

At this point, I would like to address each of these program areas, giving specific examples of how the Department is working toward a national policy of strengthening our preparedness for intentional terrorist acts, and enhancing current programs designed for the prevention or control of the unintentional introduction of harmful pathogens to our Nation's food and agriculture.

Awareness and Warning

One of USDA's key goals is to expand our surveillance and monitoring systems to provide early detection and tracing of diseases and outbreaks. In addition to expanding our systems, it is important to integrate them at a high level—identify aberrations across mission areas and sectors. Intelligence is also essential to awareness and warning so that we are knowledgeable of our enemy's intent and capabilities. We use intelligence to prioritize many surveillance and monitoring activities. As such, USDA is forging new relationships and enhancing existing relationships to improve upon our preparedness and early warning capabilities. Other efforts are ongoing in the Intelligence Community, such as the ARGUS Project, under development in the Intelligence Technology Innovation Center. Argus is a global biosurveillance system that captures and integrates massive amounts of data from many disparate sources to provide an early warning of potentially catastroophic bioevents. Our USDA programs and ARGUS will both provide feeds to the Department of Homeland Security's National Biosurveillance Detection System to help form an intregrated picture of developments in this critical area.

Animal Health Surveillance Efforts

By utilizing an integrated Federal-State system of collecting animal diagnostic samples to detect diseases at statistically significant levels, state-of-the-art laboratory networks to test samples, and analyzing data, USDA's Animal and Plant Health Inspection Service (APHIS) is able to anticipate new or emerging threats, and to quickly eliminate or contain already-identified threats. APHIS is enhancing its animal health surveillance systems by collaborating with its counterparts in the Canadian and Mexican governments. As an example, the North American Animal Health Committee, which includes experts from the United States, Canada, and Mexico, considers surveillance methods for detecting a foreign animal disease and engages in collaborative efforts to develop harmonized approaches that reduce trade disruptions.

In addition, USDA partners with HHS's Centers for Disease Control and Prevention (CDC) and FDA, the U.S. Geological Survey (National Wildlife Health Center), the Southeast Cooperative Wildlife Disease Study, the National Assembly of State Animal Health Officials, and the National Association of State Public Health Veterinarians in the Interagency Working Group for the Coordination of Zoonotic Disease Surveillance (ZDWG). The ZDWG meets by teleconference monthly to address zoonotic disease surveillance issues.

Surveillance for Avian Influenza

With growing attention to the threat of avian influenza (AI), USDA has further strengthened its already extensive efforts on the disease, including surveillance, enhanced bio-security of poultry farms, controlling the movement of birds and products, and encouraging industry practices that reduce risk. The supplemental appropriation recently enacted as part of the FY 2006 Defense Appropriations Act contains funding important to bolstering this effort.

As part of our overarching program for AI, USDA conducts ongoing surveillance in numerous ways. USDA works closely with State Agriculture Departments, universities, and industry representatives to increase surveillance and testing within the commercial poultry population, live bird markets, and migratory bird flyways. USDA is also capitalizing on our Exotic Newcastle Disease (END) surveillance program in backyard flocks by testing the END samples for AI as well.

For the last several years APHIS has conducted an outreach campaign called "Biosecurity for the Birds." The campaign places informational materials directly into the hands of commercial poultry producers, as well as those raising poultry in their backyards. All of the brochures and fact sheets are available in several languages and emphasize the need for good bio-security and disease surveillance programs to reduce the possibility of bringing any disease, not just AI, on the farm or into their back yard. USDA and CDC continue to coordinate domestic AI diagnosis and response efforts.

I would be remiss if I did not also note that the U.S. poultry industry has been extremely vigilant in its efforts to ensure that any disease is detected, contained and eradicated quickly.

Plant Pest Detection

The APHIS Pest Detection program coordinates a nationally directed survey program through the Cooperative Agricultural Pest Survey network. The program works with State and university cooperators through national, regional, and State-level committees to prioritize survey projects and provides funds for State cooperators to conduct the agreed upon surveys. The program also trains and equips State cooperators to conduct national surveys.

Offshore Pest Surveillance

APHIS maintains the Offshore Pest Information System (OPIS). OPIS is a structured, risk-focused process designed to collect, synthesize and analyze, and communicate relevant offshore agricultural pest and disease information. APHIS plant and animal health specialists located overseas monitor and track agricultural pest and disease situations for OPIS reporting. This information is used to prevent, or prepare for, the possible introduction of pests or diseases into the country.

Food Testing for Threat Agents

Several agencies within USDA collaborate to test for threat agents within the food supply. USDA's Food Safety and Inspection Service (FSIS) conducts surveillance to provide ongoing verification of food safety processes and a quick alert to contamination or an outbreak related to meat, poultry, or egg products. FSIS has expanded its longstanding regulatory sampling program to test for harmful chemical, biological, and physical hazards in these products. The proportion of samples tested for threat agents depends on the DHS threat condition. These activities, as well as FSIS' daily presence in facilities, help protect the food supply from intentional contamination.

In addition to FSIS' activities, USDA's Agricultural Research Service (ARS) is developing new tests for the detection of foodborne pathogens and toxins and is validating those tests in food matrices, which is essential to ensure that detection methods actually work in foods, not just in a test tube.

The Small Business Innovation Research program administered by USDA's Cooperative State Research, Education and Extension Service (CSREES) funds small business research and development projects, including several that address agricultural bio-security threats. Notable among these are several projects that focus on detection of food borne bio-security threats including: *Salmonella, E. coli, Listeria, Hepatitis,* and foreign contaminants.

CSREES is also funding work at the University of Minnesota to implement a national interactive web-based communication and coordination system to facilitate Federal, State and local collaboration in food protection and defense. This project, which is based on an Association of Food and Drug Officials (AFDO) prototype, leverages the resources and expertise from the

DHS-funded National Center for Food Protection and Defense, the AFDO, and the National Food and Agriculture Laboratory Committee.

Consumer Complaint Monitoring System

FSIS' Consumer Complaint Monitoring System (CCMS) is a national system to monitor and track food-related consumer complaints. The system is triaged daily by public health nurses who look for common patterns and trends that could serve as an early warning of an outbreak or an intentional contamination event. An example of building upon existing safety tools to also include security goals, CCMS is used as a real-time, early-warning system of a potential attack on our food supply. Consumers report food safety complaints by calling USDA's Meat and Poultry Hotline (1-888-MPHotline) or reporting to FSIS District Offices. The CCMS database also contains food safety complaints made to other Federal agencies, such as FDA, and State and local health departments. FSIS is working with State and local partners to include their data, thereby improving communication and data-sharing capabilities to increase detection and provide more efficient intervention.

FSIS is also building a system to support and improve the agency's incident response capabilities to natural disasters and acts of terrorism. Specifically, the Non-Routine Incident Management System (NRIMS) will provide FSIS with the ability to meet the expectations laid out in the National Response Plan (NRP). The NRIMS will maintain a unified approach to incident management that supports the incident command structure, and in the event of a major incident, will optimize collaboration with other USDA agencies, the State, and other Federal government emergency support functions.

Electronic Commodity Ordering System expansion

Similarly, USDA's Food and Nutrition Service's (FNS) Electronic Commodity Ordering System (ECOS) program also builds on an existing program to enhance safety or security. Adapting the ECOS to include a commodity food safety complaint component is the first step in implementing a rapid alert and notification system to reach State and local commodity recipients with up-to-the-minute food safety information. This change allows local schools to report defective foods in a timely manner and enables USDA to see trends in complaints and 'connect the dots' should an intentional contamination appear in different places at the same time.

Laboratory Networks

Detecting pests, disease outbreaks or contamination quickly enables us to determine the origin, respond, and mobilize sooner, which reduces the impact of an event. Therefore, our laboratories are important surveillance tools. To enhance our detection ability and our response capabilities, we have established national networks of Federal and State laboratories with the capacity to test animal, plant, and food samples for threat agents in the event of a terrorist attack. USDA is also part of a consortium of laboratory networks including animal, plant, food, public health, defense and environmental interests.

CSREES works closely with APHIS and other partners to coordinate the National Animal Health Laboratory Network (NAHLN) and the National Plant Diagnostic Network (NPDN). Both networks have the ability to respond to biological agents whether intentionally or unintentionally introduced.

The NAHLN is a joint effort among CSREES, APHIS's Veterinary Services and the American Association of Veterinary Laboratory Diagnosticians to create a functional national network from existing animal diagnostic laboratories. At the end of FY 2005, the NAHLN consisted of 49 State and university laboratories in 41 States. Surveillance for transmissible spongiform encephalopathies (including bovine spongiform encephalopathy, chronic wasting disease and scrapie) is occurring in 26 NAHLN labs. A surveillance plan for classical swine fever (CSF) was developed and is being implemented in States with a high risk for introduction of CSF. Twelve NAHLN labs will be testing CSF samples and 18 other NAHLN labs will assist with sample collection and processing. Assays for AI and END have been deployed in 38 labs throughout the United States. The support and development of the NAHLN network is a strong weapon against bio- and agro-terrorism.

Similar to NAHLN, the NPDN is a joint effort between CSREES and APHIS Plant Protection and Quarantine to create a national network from the existing plant diagnostic laboratories. Recently, the NPDN has been an important component of the soybean rust (SBR) coordinated framework that provides timely information about SBR, thereby saving soybean growers unnecessary fungicide applications.

USDA and HHS have developed an integrated laboratory network, the Food Emergency Response Network (FERN). FERN is a collaborative effort that includes Federal and State laboratories, coordinated jointly by FSIS and FDA to provide ongoing surveillance and monitoring, as well as a rapid and local response when necessary. Laboratories participating in

FERN are responsible for detecting and identifying biological, chemical and radiological agents in food.

To enhance surveillance capabilities, laboratory networks from a variety of Federal Departments have agreed to work cooperatively in an Integrated Consortium of Laboratory Networks. All have signed onto a Memorandum of Understanding to communicate and cooperate by sharing capabilities, policies, procedures and approaches for handling laboratory analysis for routine surveillance and surge capacity during national emergencies. The consortium also seeks to reduce redundancies among laboratories, identify holes in laboratory capabilities and to find solutions to managing these identified issues in the future. Together our network of resources work to enhance detection of – and enables a rapid and sufficient response to – food, animal, plant and human health emergencies.

Coordination with the Intelligence Community and Law Enforcement

USDA is expanding its partnerships to include nontraditional partners such as intelligence community members and law enforcement agencies. One way to develop strong relationships is to work side-by-side with these entities. A senior intelligence advisor is assigned to USDA's Homeland Security Office and works primarily on information sharing between the intelligence community and USDA, bolstering that vital connection. In addition, FSIS has staff assigned to work closely with the National Counter-Terrorism Center, and APHIS and FSIS attend working level staff meetings of the interagency Ag-Intel Working Group to exchange information on potential threats to the sector.

Vulnerability Assessments

Vulnerability assessments play a key role in helping us to determine and implement the most effective countermeasures to prevent a terrorist attack on our sector.

Interagency Site Assistance Visits

USDA is partnering with the FBI, FDA, and DHS on the Strategic Partnership Program for Agroterrorism, a program to visit a variety of industries within the sector to validate previously conducted vulnerability assessments, initiate new assessments, and foster improved relations among industry, State agriculture and health officials, local law enforcement, FBI officials, and USDA and FDA field staff. Findings will also be used in considering mitigation strategies and to populate DHS databases. To ensure appropriate government and industry participation, the effort is synchronized via the Sector Coordinating Councils. As examples, USDA has partnered with the grain industry in Louisiana; will begin working with North Carolina to examine vulnerabilities in central kitchens preparing food for the National School Lunch Program; and, plans to work with the egg industry in Pennsylvania. USDA will coordinate with FDA to work with industries that are jointly regulated.

FSIS has completed eight vulnerability assessments for selected domestic and imported food products. APHIS has completed four assessments for selected agricultural production industries. Both agencies have provided technical expertise concerning threat scenarios and the application of the CARVER + Shock assessment tool to industries conducting their own assessments. These assessments help industry learn more about their vulnerabilities and provided a forum to consider mitigation strategies. USDA agencies have worked with private sector entities including the National Pork Producers Board, the Texas Cattle Feeders Association, the American Meat Institute, National Turkey Federation, United Egg Association, and Kraft Foods. USDA plans to formalize its agencies' outreach and assistance via the Interagency Site Assistance Visits.

CSREES is also working to further enhance food security assessments. It is funding multi-disciplinary research at Rutgers University to better understand public awareness, perceptions and reactions and to identify the key information needs in the event of a food security incident. The project includes research to facilitate rapid communications, investigations, trace back and containment. In addition, the project is providing fundamental training to several graduate students who are developing critical food bio-security research expertise.

CSREES also supports a Multi-State Research Committee specifically to assess the potential impacts of bioterrorism on crop production and trade. As an illustration of the dual utility of this effort, one of the project members had analyzed the probable impacts of a hypothesized disruption of the grain handling capacity at the Port of New Orleans well before the devastation of Hurricane Katrina occurred. As a result, valuable information was available when this disaster occurred.

Mitigation Strategies

Early awareness enables a more effective mitigation strategy. USDA has the goals of developing animal tracking systems and expanding screening and inspection procedures, so that we may quickly respond to an attack or naturally occurring incident.

National Animal Identification System

The implementation of a national animal identification and tracking system (NAIS) is a priority for USDA. NAIS will enhance the speed and efficiency of disease trace backs by standardizing animal movement recordkeeping and using newer technologies. Upon full NAIS implementation, our goal is to be able trace the movements of all exposed or infected animals entered in the NAIS within 48 hours of a disease diagnosis.

USDA, State agencies, Tribal Nations, and agricultural stakeholders are making strides toward full NAIS implementation. Currently, all the States, 5 Tribes, and 2 U.S. Territories are capable of registering premises according to USDA standards, and more than 150,000 locations have received a nationally unique premises identification number. USDA is also preparing to begin allocating animal identification numbers in the near future. Stakeholder comments on the NAIS strategic plan and draft program standards indicate strong support for NAIS. While views among stakeholders vary, a majority of producers who commented on these documents favor a system that allows animal movement data to be privately held. On August 30, 2005, USDA announced the development of a public/private partnership that enables the private sector to maintain animal movement data as part of the NAIS. Efforts continue to register premises and establish a solid framework for construction of an effective NAIS in the years to come.

Targeted Screening and Inspection of Imported Food

FSIS ensures that meat, poultry and egg products imported into the United States are produced in a system that provides the same level of protection against food safety hazards as the domestic system. In addition, when each meat, poultry and egg product shipment enters the country under the authority of U.S. Customs and Border Protection, FSIS import inspectors verify that each lot is properly certified, examine each lot for general condition and labeling and conduct reinspection as appropriate. To aid in ensuring the importation of safe and wholesome foods, FSIS established Import Surveillance Liaison Officers (ISLOs), and assigned them to port cities on the perimeter of the United States. FSIS has 22 ISLOs who conduct a broad range of surveillance activities at seaport, land entry points along the Canadian and Mexican borders, and air terminals across the United States. Their focus is beyond the inspection areas where import inspectors work and includes docks, refrigeration and storage areas, and they coordinate with other agencies concerned with the safety and security of imported foods.

Minimizing the Impact of Soybean Rust on the Agriculture Industry

In November 2004, USDA published its Strategic Plan to Minimize the Impact of the Introduction and Establishment of Soybean Rust on Soybean Production in the United States, which involves a coordinated effort of Federal and State agencies, industry, growers and crop consultants to enable recovery from the introduction of SBR. On March 15, 2005, the USDA SBR website was launched to provide a one-stop source for information, including an early warning system on detections and appropriate control recommendations. The ARS is conducting research to develop effective fungicide treatments to mitigate any outbreaks or prevent them from occurring. In addition, ARS scientists are actively involved in the identification and development of disease resistant germplasm.

Through a partnership agreement between CSREES, USDA's Risk Management Agency and North Carolina State University, additional crop-pest combinations, including some of concern in Pennsylvania, will be added to the program this year. This partnership will build on a CSREES National Research Initiative (NRI) research grant to Pennsylvania State University to develop a new Pest Information Platform for Extension and on the capabilities of Zedex, a private meteorological firm in Bellefonte, Pennsylvania.

Preventing Animal and Plant Pests and Diseases from Entering the United States

USDA remains committed to maintaining a strong relationship with DHS and working cooperatively to ensure the continued success of agricultural inspection operations at all U.S. ports of entry. For example, APHIS is working with DHS's Customs and Border Protection to establish a quality assurance program for agricultural inspections. This program will ensure the quality and thoroughness of inspections and further facilitate communications.

USDA and DHS are also cooperating on new technologies to enhance border inspection efforts, including development of an automated inspection system to screen manifests electronically and target high-risk cargo; remote digital imaging to quickly identify pests on imported items; and, a nationwide database of regulation violators.

Response Planning and Recovery

In the event of an attack or unintentional contamination or outbreak, it is important that we all know our respective roles and responsibilities. The NRP, issued by DHS, is integral to ensuring coordinated incident responses.

National Response Plan Implementation

USDA is identifying and preparing revisions to existing regulations, policies and guidance to assure compliance with the NRP. USDA has developed and is delivering NRP training courses for USDA employees and stakeholders.

Under a cooperative agreement, USDA, DHS, FDA and the National Association of State Departments of Agriculture formed a working group to develop the Food and Agriculture Annex to the NRP and the interagency Food and Agriculture Response Plan that will implement the annex. To date, the working group has submitted the Annex for interagency review; finalized the template for the Plan; and, is in the process of pilot testing the template.

USDA is also sharpening its readiness for the Incident Command System. This is a part of a command approach that gives Federal, State and local governments a unified strategy for working together to prepare for, respond to, and recover from domestic incidents.
APHIS currently implements the Incident Command System when responding to animal and plant health incidents. In conjunction with our State colleagues, there are State-level emergency response teams on standby. These teams will typically be on site within 24 hours of the detection of a plant pest or animal disease.

FSIS is also developing a series of exercises to take place across the country over the next three years to prepare to respond in the event of an incident involving the intentional contamination of the food supply. Stakeholders encouraged to participate in the exercises include Federal, State and regional food safety, public health, emergency management and law enforcement agencies and affected industries. One objective of the exercise is to test the ability of the agencies to respond within an incident command system.

National Plant Disease Recovery System

USDA is working with Federal agencies, State and local governments and the private sector to develop a system, the National Plant Disease Recovery System (NPDRS), capable of responding to a high consequence plant disease. ARS has assumed leadership of this effort and initiated a roadmap for implementation. NPDRS will implement sufficient control measures and develop resistant seed varieties for economically important crops.

Disposal and Decontamination

USDA also recognizes the need to look beyond the initial response and be prepared to assist in the recovery of the industries that constitute our critical infrastructure. In the event of an intentional contamination of the food supply, there is the potential that large volumes of food products will need to be disposed of and food processing facilities will need to be decontaminated. FSIS has developed, in conjunction with FDA and the Environmental Protection Agency, a document that provides guidance as to the roles and responsibilities of various agencies in disposal and decontamination activities, options that would be available for disposal of contaminated food products and decontamination of facilities depending upon the type of agent used in the attack and a list of important contact numbers when dealing with disposal and decontamination issues. FSIS is also working with and encouraging industry and

States to start preparing and planning for the disposal of food, which might be classified as a hazardous waste, in advance of an event.

National Veterinary Stockpile

APHIS administers the National Veterinary Stockpile (NVS) for specific, high threat foreign animal diseases. It is capable of maintaining vaccines for use in the United States in the event of a significant foreign animal disease outbreak. APHIS will use the NVS to consider and obtain "ready-to-use" vaccine products. The goal is for NVS to become one component of an overall response planning and recovery effort to provide the best possible protection against an attack on our food and agriculture system. USDA is working with DHS, FDA, and the Environmental Protection Agency in ensuring that adequate supplies of animal vaccines, antivirals, and therapeutics have been stockpiled to appropriately respond to the most damaging animal diseases affecting human health and the economy.

APHIS has awarded a five-year contract to Fort Dodge Animal Health to develop an AI vaccine antigen bank for poultry that will house enough frozen antigen to produce up to 10 million doses of vaccine for a variety of AI subtypes. In the event of a high pathogenicity AI outbreak, vaccination could be used to create barriers against further spread and assist with our overall control and eradication measures.

Coordination with CDC

USDA and CDC continue to coordinate interagency response efforts with respect to food safety and security, especially at the animal-human interface. Two permanent veterinary personnel staff from USDA have been posted to CDC. These liaison positions and established relationships have served the agencies well in times of crisis, most recently during the avian influenza outbreak in Southeast Asia. The agencies offered technical assistance to one another and shared information on the status of affected countries, import bans, and the prevention of occupational exposures.

Outreach and Professional Development

Outreach and professional development provide fundamental underpinnings for USDA bio-security preparedness by assuring that Federal personnel and their State, local and private sector cooperators possess the core knowledge and abilities that will be necessary. It is equally important to have outreach capacity to educate and inform the citizens directly affected by bio-security threats and incidents.

CSREES provides the leadership and funding for the Extension Disaster Education Network (EDEN) which is a collaborative multi-State effort linking Extension specialists across the country. EDEN's mission is to share education resources to reduce the impact of natural and man-made disasters. This mission is carried out through interdisciplinary and multi-State research and education programs addressing disaster mitigation, preparation, response and recovery. EDEN works by linking with Federal, State and local agencies and organizations in anticipation of future disaster education needs to prepare communications and deliver timely information that meets local needs. EDEN provides a national clearing house of pertinent information and a network of disaster recovery experts. This network in turn provides information and expertise to the local Extension workers who have direct, preexisting relationships with local growers and regulatory officials.

APHIS is educating producers and veterinarians on bio-security so that they can be prepared to identify clinical signs of infectious diseases. APHIS worked with DHS' Office for Domestic Preparedness to develop an Agriculture Emergency Response Training (AgERT) course in Anniston, Alabama. AgERT prepares APHIS employees to serve as emergency responders and provides traditional first responders with training for managing emergencies in agricultural settings. The course is targeted to responding to emergencies in an agricultural setting, but its instruction covers the use of personal protective equipment in all manner of emergencies, including chemical, radiological, and biological emergencies. Because individual training is not always possible, USDA also uses CD and web-based training. They are particularly effective mechanisms when the target audiences are widespread and in rural areas where traveling to a training site is difficult.

FSIS is also working with industry organizations to actively seek out opportunities to encourage adoption of food defense activities, especially the use of vulnerability assessments and model food defense plans. To educate industry on actions they may take to increase security at their facility or within the system, and to continue to educate consumers on the importance of safe food handling practices that will help reduce the impact of intentionally contaminated foods, FSIS issued three sets of voluntary guidance documents:

- "Food Security Guidelines for Food Processors," which targets slaughter and processing plants. It helps establishments identify ways to strengthen their protection against intentional contamination.
- "Safety and Security Guidelines for the Transportation and Distribution of Meat, Poultry and Egg Products," which helps facilities and shippers that process or transport meat, poultry and egg products identify potential vulnerabilities in their own operations and address them.
- "Food Safety and Food Security: What Consumers Need to Know," which outlines practical information for consumers about safe food handling practices, foodborne illness, product recalls, keeping foods safe during an emergency and reporting suspected instances of food tampering.

Using the guidance materials, FSIS prepared a self-assessment checklist for industry to use to assess the security of their operations. In addition, to encourage this kind of voluntary action in protecting the food supply against all threats, FSIS has released four model food defense plans for egg processing facilities, meat and poultry processing facilities, slaughterhouses, and import facilities. These plans are based on vulnerability assessments and checklists and are geared to serve as models to assist industry in developing their own facility-specific food defense plans. To help industry adopt the plans, USDA is reaching out to all target establishments with a specific emphasis on smaller companies that might not have the resources to develop their own independent food defense plans. As part of its outreach program, FSIS is providing training tailored to small and very small plants to encourage industry to adopt food defense plans. While these guidelines are voluntary, FSIS strongly urges all establishments operating under Federal and State inspection programs to incorporate these security procedures.

Integral to the implementation of food defense plans by industry is the understanding of what are the "best practices" to securing a particular establishment. FSIS is interested in working with industry to identify these "best practices." Where there is a deficiency in available options, FSIS intends to encourage the development of new technologies to assist industry to better secure their plants.

Food Security Awareness Training

To foster an awareness of the need to prevent and respond to food security threats with partners in food safety, FSIS conducted fourteen food security awareness sessions for food defense partners at locations throughout the United States. The sessions provided information about the CARVER + Shock method of conducting vulnerability assessments and general food security awareness topics. There were 1,040 participants in the sessions from a variety of Federal, State and local organizations, including FDA, State and county health departments, and School Lunch Program administrators.

Agriculture Transportation Security Guidance

USDA and the American Trucking Association developed a voluntary security guidebook and risk assessment tool for use by truck company owners and drivers to enhance security from external threats, including terrorism, and to protect trucking facilities and vehicles.

Externships and Fellowships

The field of food and agriculture security is relatively new, and therefore, has few experts. Ensuring a competent and robust workforce of the future is important to the security of the sector in the long term. USDA has a number of initiatives to encourage study in this field. APHIS is collaborating with Veterinary Schools to establish externship programs between senior veterinary students and various units within the agency's Veterinary Services program. FSIS currently employs several post-graduate fellows who are engaged full-time in bioterrorism and vulnerability assessment activities.

Research and Development

Ultimately scientific research and development is the basis of our ability to prepare for, detect, respond to and recover from bio-security incidents. Within USDA, multiple scientific disciplines including biological, physical, social, information technology and communications are contributing to our preparedness. The agencies of USDA's Research, Education and Economics mission area are all contributing to this effort in a coordinated and collaborative manner.

ARS is USDA's intramural research arm. ARS conducts research to support the needs of the action agencies of the department and the Federal government. In addition, it conducts research in partnership with academia and industry, as well as other governmental agencies at the Federal and State levels to carry out its mission through a number of highly coordinated national programs. ARS research related to homeland security falls into three broad areas: 1) plant protection, 2) animal protection, and 3) food safety. Included in these activities are the development of new diagnostic and detection methods, preventative activities such as the development of new vaccines and risk reduction intervention strategies, and the mitigation of a variety of threats.

ARS has built collaborations with States on a number of bio-security research issues. With supplemental appropriations received in 2002 to develop rapid tests for threat agents, ARS shared those funds with a number of universities to accomplish its goals. One particular example was the construction of a plant threat agent database at Pennsylvania State University. By collaborating on research, ARS can share efforts in developing and disseminating bio-security technologies. All of the rapid tests that come from this research is being provided through APHIS and CSREES and directly to members of the Plant and Animal Health networks, which are State based laboratories.

ARS and the land grant universities are the main contributors to the national germplasm system. That system will play a key role in the NPDRS. ARS is presently on the campuses of most land grant universities and cooperates with those schools to accomplish its research goals. This cooperative relationship assures that State universities are a continuing part of the initiatives of ARS research.

One example relevant to Pennsylvania's agricultural industry is the AI research funded by the NRI. This NRI Coordinated Agricultural Project entitled "Prevention and Control of Avian Influenza in the United States" includes universities in seventeen States, including Pennsylvania State University and the University of Pennsylvania. Working together in a coordinated manner they are developing diagnostic tests and vaccines as well as monitoring live bird markets and wild waterfowl flyways to enable early detection and rapid response.

USDA is constructing a world-class animal disease bio-containment facility for research and diagnostics in Ames, Iowa that will house the Centers for Animal Health, which are the National Animal Disease Center, the NVS Laboratory and the Center for Veterinary Biologics. In this context, USDA is developing a comprehensive suite of rapid diagnostic tests to detect and identify pathogens within hours that pose the greatest threat to U.S. livestock.

Regional Dairy Quality Management Alliance

The ARS collaboration with the Regional Dairy Quality Management Alliance (RDQMA) program extends the concept of epidemiology to monitor pathogen presence and help dairy producers assure product safety. Microbial pathogens are a continuing concern in dairy production due to their potential impact on animal health, milk production and safety, and economics. ARS collaborates with scientists from three universities in quarterly farm visits assessing herd management practices and bio-security, and in sampling and laboratory evaluation of animal and food safety pathogens, including animal diseases such as Johne's Disease and food safety pathogens such as Salmonellosis. This ARS and university collaboration, including activities in Pennsylvania, will provide the data to validate and implement best management practices for commercial dairy cattle herds that are a valuable component of bio-security programs. RDQMA is building and maintaining a database of all the records from the annual and quarterly surveys, as well as individual cow information. This national database of well-characterized animals can be used for future research as well as a basis for comparison to evaluate and mitigate bio-security concerns.

USDA Research on Crop Disease Detection and Food Defense

ARS has worked closely with other USDA agencies and other Departments to address research and development needs resulting from vulnerability assessment findings. Methods were successfully developed to detect B. anthracis in milk on the farm, during transport/handling and at the processing plant to assure bio-security.

ARS scientists working on crop diseases have developed rapid tests for plant threat agents. These highly sensitive and accurate tests provide diagnosticians with an accurate means to detect pathogens as part of a national surveillance system. The rapid test for SBR played an important role in the detection of SBR, in following its spread, and in the application of technology to reduce the impact of SBR.

Future research will ensure that disease resistant varieties of plants and crops are continuously developed and made available to farmers and producers. ARS will develop real-time, field deployable surveillance and detection methodologies, coordinate the validation process, and determine how the environment affects the establishment, spread and persistence of a threat in an agricultural context.

Modeling and Mapping Development

USDA's Economic Research Service (ERS) developed the Geo-Spatial Economic Analysis (GSEA) System to serve as a platform for collaborative analysis of the economic consequences of natural and man-made threats to the food and agricultural industries. The GSEA system can be used to examine the economic impact of events that disrupt the production, processing, distribution or consumption of food and agricultural products. The key to this system is the ability to leverage existing expertise and analytical capacity at ERS by identifying and filling data gaps, integrating the results of dissimilar economic analyses and developing interfaces with plant and animal epidemiology models.

Coordinated Budget

HSPD-9 directed a coordinated budget submission by USDA and HHS for food system defense to ensure collaboration during program and budget planning. USDA coordinated with HHS to submit the Food and Agriculture Defense Initiative for FYs 2005 and 2006. We continue to work together on the FY 2007 initiative.

Conclusion

I am sure we are united in the belief that bio-security for food and agriculture is a paramount issue that must be addressed with sound science and policy decisions. The Department looks forward to working with the Committee in continuing to develop programs and initiatives to help enhance the security of the nation's agriculture and food systems through collaborative efforts between Federal and State institutions. Mr. Chairman, that concludes my remarks. I would now be pleased to take any questions.

Testimony
Before the Subcommittee on Research, Nutrition, and General Legislation
Committee on Agriculture, Nutrition, and Forestry
United States Senate

The Role of HHS in Agricultural Counterterrorism Activities

Statement of

Alex M. Azar, J.D.

Deputy Secretary
U.S. Department of Health and Human Services

For Release on Delivery
Expected at 1:15pm
Monday, January 9, 2006
Harrisburg, Pennsylvania

INTRODUCTION

Good afternoon, Chairman Santorum and Members of the Subcommittee. I am Alex Azar, Deputy Secretary of the Department of Health and Human Services (HHS or the Department). I am pleased to be here today with my colleague, Charles Conner from the U.S. Department of Agriculture (USDA), and with Mr. Dennis Wolff of the Pennsylvania Department of Agriculture. HHS appreciates the opportunity to discuss our food counterterrorism activities.

The events of September and October 2001 made it very clear that terrorism - indeed bioterrorism - is a serious threat to our Nation and the world. The Bush Administration and Congress responded forcefully to this threat by providing funding to strengthen our medical and public health capacities to protect our citizens from future attacks.

A great deal has been done in the past few years to enhance the safety and security of the food supply in the United States. Within HHS, the Food and Drug Administration (FDA) has worked with food safety agencies, as well as with law enforcement and intelligence-gathering agencies, and with industry to significantly strengthen the nation's food safety and security system across the entire distribution chain, from farm to table, to better protect our food supply against deliberate and accidental threats. This cooperation has resulted in greater awareness of vulnerabilities, the creation of more effective prevention programs, new surveillance systems, and faster foodborne illness outbreak response capabilities. HHS's Centers for Disease Control and Prevention (CDC) maintains national surveillance for specific infections and for outbreaks of foodborne illnesses, supports states in investigating and controlling outbreaks, and maintains cross-communication with FDA and USDA.

Food safety and food defense continue to be top priorities for this Administration. A terrorist attack on the food supply could have both severe public health and economic consequences, while damaging the public's confidence in the food we eat. The changes in food safety and defense that we have been implementing in the last few years are the most fundamental enhancements in our food safety and defense activities in many years.

In my testimony today, I will first briefly describe HHS' overall role in counterterrorism activities. Then, I will discuss our collaborative activities with our food safety and defense partners. I will also describe some of FDA's counterterrorism activities to enhance protection of the food supply. Finally, I will briefly discuss some of our efforts with regard to avian influenza.

HHS' ROLE IN COUNTERTERRORISM ACTIVITIES

Under the President's National Response Plan, HHS leads federal public health efforts to ensure an integrated and focused national effort to anticipate and respond to emerging biological and other weapons threats. HHS is also the principal federal agency responsible for coordinating all Federal-level assets activated to support and augment the state and local medical and public health response to mass casualty events.

Principally through HHS's Centers for Disease Control and Prevention (CDC) and Health Resources and Services Administration (HRSA), funds have been provided to States and localities to upgrade infectious disease surveillance and investigation, enhance the readiness of hospitals and the health care system to deal with large numbers of casualties, expand public health laboratory and communications capacities and improve connectivity between hospitals, and city, local and state health departments to enhance disease reporting.

CDC also operates HHS's Strategic National Stockpile (SNS), which contains large quantities of medicine and medical supplies to protect the American public if there is a public health emergency severe enough to cause local supplies to run out. Once Federal and local authorities agree that the SNS is needed, medicines will be delivered to any state in the U.S. within 12 hours. Consequently, each state is now required to develop plans to receive and distribute SNS medicine and medical supplies to local communities as quickly as possible in the event of a deployment.

HHS's National Institutes of Health (NIH) is assigned the lead role in the development of medical countermeasures to biological attack, and in the conduct of research concerning potential agents of bioterrorism that directly affect human health. The National Institute of Allergy and Infectious Diseases (NIAID) is the NIH institute with primary responsibility for carrying out this assignment.

To further encourage the development of new medical countermeasures against chemical, biological, radiological and nuclear agents and to speed their delivery and use should there be an attack, President Bush, in his 2003 State of the Union address proposed and Congress subsequently enacted Project BioShield to assure developers of medical countermeasures that funds would be available to purchase these critical products for use to protect our citizens. Project Bioshield is operated out of the HHS Office of Public Health Emergency Preparedness, which also coordinates the HHS-wide emergency preparedness activities and serves as the principal point of contact at HHS for other Federal agencies and Departments.

FDA is the Federal agency that regulates everything we eat except for meat, poultry, and processed egg products, which are regulated by our partners at USDA. FDA's responsibility extends to live food animals and animal feed. FDA also is responsible for ensuring that human drugs, human biological products, medical devices, and radiological products as well as veterinary drugs are safe and effective and that cosmetics are safe. In addition, FDA is responsible for ensuring that the health consequences of foods and medicines are accurately and honestly represented to the public, so that they can be used as effectively as possible to protect and improve the public health.

FDA's primary mission is to protect the public health. Ensuring that FDA-regulated products are safe and secure is a vital part of that mission. While performing our mission, we play a central and a leadership role in the nation's defense against terrorism. First, terrorists could use an FDA-regulated product, such as food, as a vehicle to introduce biological, chemical, or radiological agents into the U.S. stream of commerce. Second, FDA-regulated products, such as human drugs, vaccines, tissues, blood, blood products, and medical devices, as well as veterinary drugs, will play a central role in preventing or responding to human and/or animal health concerns created by an act of terrorism. It is HHS's goal, with FDA working closely with other HHS agencies and other Federal agencies, and with state and local governments, industry, and the public, to reduce the likelihood that an FDA-regulated product could be used to poison or otherwise terrorize Americans. We also help ensure that the nation's public health system is prepared to deter a potential threat and is ready to respond to an act of terrorism.

By way of background, although FDA has the lead responsibility within HHS for ensuring the safety of food products, the CDC has an important complementary public health role. As the lead Federal agency for conducting disease surveillance, CDC monitors the occurrence of illness in the U.S. attributable to the entire food supply. The disease surveillance systems coordinated by CDC provide an essential early-information network to detect dangers in the food supply and to reduce foodborne illness. In addition, these systems can be used to indicate new or changing patterns of foodborne illness. Because CDC detects and investigates outbreaks of foodborne illness through its networks, CDC is able to alert FDA and USDA about implicated food products associated with foodborne illness and works closely with the agencies to take protective public health action. In keeping with its agency mission, CDC also identifies, evaluates, and provides expert scientific opinion on the effectiveness of foodborne disease prevention strategies.

COLLABORATION WITH FOOD SAFETY AND FOOD DEFENSE PARTNERS

In its food safety and defense efforts, FDA has many partners – Federal, state and local agencies, academia, and industry. FDA is working closely with our Federal partners such as USDA, DHS, the Homeland Security Council at the White House, the Department of State, the Central Intelligence Agency (CIA), and the FBI to have the best information possible and to be prepared to act as needed. I also want to emphasize FDA's close working relationships with its sister public health agency, CDC, with Customs and Border Protection (CBP) in DHS, and with USDA's Food Safety and Inspection Service (FSIS), FDA's counterpart agency responsible for meat, poultry, and processed egg products. Some other Federal partners include USDA's Animal and Plant Health Inspection

Service (APHIS), USDA's Foreign Agriculture Service, USDA's Agricultural Research Service, USDA's Food and Nutrition Service, Department of the Army Veterinary Services Activity, the Environmental Protection Agency (EPA), and the Department of Treasury's Alcohol and Tobacco Tax and Trade Bureau.

FDA's activities in public health defense are coordinated through the HHS Secretary's Operations Center. This relationship facilitates communication among all HHS Operating Divisions, the Department, and other Federal agencies and departments, including DHS. FDA also has worked closely with the Interagency Food Working Group of the White House Homeland Security Council on three initiatives – development of a national network of food laboratories, identification of vulnerabilities and subsequent mitigations for commodities of concern, and the development of a national incident management system. In addition, FDA worked in partnership with the U.S. Environmental Protection Agency, USDA, DHS, and the Department of Defense to describe general Federal roles responsibilities for decontamination and disposal in response to animals, crop, and food incidents.

In addition, FDA's Office of Criminal Investigations (OCI) maintains professional relationships with domestic and foreign law enforcement agencies to receive and act on any information regarding the intentional contamination of FDA-regulated products. OCI has a specialized staff with the clearances, capabilities, and backgrounds to analyze information from law enforcement and intelligence community agencies and to assist those agencies in conducting terrorism-related threat assessments involving FDA-regulated products. OCI serves as FDA's liaison with the intelligence community (CIA, FBI, Defense Intelligence Agency, National Counter-Terrorism Center, and others). In this liaison capacity, OCI maintains relationships and provides expert assistance on scientific, technical, or criminal issues to specialized units within those agencies. OCI field agents serve on selected Joint Terrorism Task Forces around the country and on other multi-agency counterterrorism task forces. OCI agents actively participate in daily briefings at the FBI-led National Joint Terrorism Task Force and at the Department of Homeland Security Information Analysis Infrastructure Protection. FDA also has an OCI agent assigned on a full-time basis to Interpol's office in Washington, D.C. OCI's coordination of the agency's criminal investigative matters, including those that relate to potential acts of terrorism, help to prevent, deter, detect, and interdict a terrorist attack on FDA-regulated products.

FDA is working closely with DHS and other Federal agencies to implement the President's Homeland Security Presidential Directives (HSPDs). The Secretary of DHS is responsible for coordinating the overall national effort to enhance the protection of the critical infrastructure and key resources of the nation, including food and agriculture defense. The President has issued HSPD-7,-8, and–9, which identify critical infrastructures, improve response planning, and establish a national policy to defend the agriculture and food systems against terrorist attacks, major disasters, and other emergencies. HSPD-9 calls for the development of a National Veterinary Stockpile (NVS). FDA and CDC participate in NVS Steering Committee activities.

The HHS and USDA Secretaries or their designees exercise key responsibilities as food sector-specific agencies. DHS serves as the coordinator of the Food and Agriculture Sector within the Government Coordination Council (GCC). The GCC is charged with coordinating agriculture and food security strategies and activities, policy, and communication across government and between the government and the sector. In addition, the Council plays a coordinating role with the public health and clinical issues resulting from a terrorist act involving the food supply.

Within the GCC, HHS and USDA serve as co-leads for the food sector, and USDA serves as the lead for the agriculture sector. The Food and Agriculture Sector is a public-private partnership that combines expertise from several Federal agencies (FDA, USDA, EPA, Department of Defense [DoD], Department of Commerce, Department of the Interior, and the Department of Justice) as well as that of state and local officials (representing agriculture, public health, and veterinary services), and the private sector (more than 100 trade associations and individual firms participate). As part of the HSPD-7 National Infrastructure Protection Plan (NIPP) development, FDA and USDA have drafted sector-specific plans, which will be revised after obtaining additional input from states and the private sector. Using these plans as components, DHS has formulated the Interim NIPP for all sectors. With the close working relationship of FDA and USDA and the other government and industry collaborators, the Food and Agriculture Sector activities to protect critical infrastructure have set the organizational and operational standard for other critical infrastructure sectors. DHS has applauded the Food and Agriculture Sector's organizational structure, consensus building, and the steps it has taken to improve food defense.

FDA also is working closely with our state partners to enhance food defense. For example, during the fall of 2004, FDA issued the Food Security Surveillance Assignment to FDA field personnel and participating state authorities to conduct food defense-related inspections, reconciliation examinations, and collections and analyses of samples of food products that have an elevated risk for intentional contamination. The purpose of this assignment was to deter intentional contamination of food through heightened and targeted preventive activities and to identify and address any gaps in the system for responding to a period of increased food security risk. This assignment enhanced both FDA's and our state counterparts' preparedness for a future threat involving an FDA-regulated product. Since that time, FDA has issued and completed three additional assignments to further integrate our food defense activities into our food safety work.

In addition, FDA and CDC have been collaborating with a Council of Association Presidents to develop a nationwide food defense awareness training program.

This Council, which consists of ten of the major state and local public health and regulatory professional associations, has an outreach capability to reach virtually all state and local public health officials. The training program will help raise food defense awareness at the local, state, and Federal levels.

Now, I would like to describe some of FDA's other counterterrorism activities.

IMPORTS

In Fiscal Year (FY) 2005, FDA had the challenge of reviewing and/or inspecting more than 8.6 million imported food line entries. In FY 2006, we expect 10 million imported food line entries. In recent years, FDA's presence has expanded at ports of entry, increased surveillance of imported foods, focused on high-risk domestic inspections, and enhanced our laboratory analysis capacity. To manage the ever-increasing volume of imported food shipments, FDA is working to utilize more risk-management strategies in the review of foods that are being imported or offered for import into the United States. Currently, working with information submitted either through CBP's electronic systems used for import entries or through FDA's new Internet-based Prior Notice System Interface, FDA screens shipments electronically before they arrive in the U.S. to determine if the shipment meets identified criteria for physical examination or sampling and analysis or warrants other review by FDA personnel. This electronic screening allows FDA to better determine how to deploy our limited physical inspection resources at the border on what appear to be higher-risk food shipments while allowing lower-risk shipments to be processed in accordance with traditional import procedures after the electronic screening.

IMPLEMENTATION OF THE PUBLIC HEALTH SECURITY AND BIOTERRORISM PREPAREDNESS AND RESPONSE ACT OF 2002 (BIOTERRORISM ACT)

Subtitle A of Title III of the Bioterrorism Act provided the Secretary of Health and Human Services with new authorities to protect the nation's food supply against the threat of intentional contamination and other food-related emergencies. This legislation represents the most fundamental enhancement to our food safety authorities in many years. These additional authorities improve our ability to act quickly in responding to a threatened or actual terrorist attack, as well as other food-related emergencies. Since this legislation was signed into law three years ago, FDA has been working hard to implement this law effectively and efficiently. Throughout this process, FDA has enjoyed close cooperation from our colleagues at CBP. I would now like to describe FDA's actions to implement several of the provisions in the Bioterrorism Act.

Registration of Food Facilities

Section 305 of the Bioterrorism Act requires registration of foreign and domestic food facilities that manufacture, process, pack, or hold food for consumption by humans or animals in the U.S. Thanks to this provision, FDA has, for the first time, a roster of foreign and domestic food facilities that provide food for

American consumers. In the event of a potential or actual terrorist incident or an outbreak of foodborne illness, the registration information will help FDA to quickly identify, locate, and notify the facilities that may be affected.

On October 10, 2003, FDA and CBP jointly published an interim final rule to implement the registration requirement, which became effective on December 12, 2003, as required by the Bioterrorism Act. The registration interim final rule was effective immediately but provided an opportunity for public comment on specific issues. On October 3, 2005, FDA issued the Registration of Food Facilities Final Rule, which affirmed the requirements initially set forth in the interim final rule. As of December 2, 2005, approximately 271,000 facilities have registered with FDA. This includes about 116,000 domestic and about 155,000 foreign facilities.

Prior Notice of Imported Food Shipments

Section 307 of the Bioterrorism Act requires the submission to FDA of prior notice of food, including animal feed, that is imported or offered for import into the U.S. This advance information enables FDA, working closely with CBP, to more effectively target inspections of food at the border at the time of arrival to ensure the safety and security of imported foods. On October 10, 2003, FDA and CBP jointly published an interim final rule to implement this provision. The interim final rule provided stakeholders an additional opportunity to comment on all provisions of the interim final rule for almost six months while the rule took effect on December 12, 2003, as required by the Bioterrorism Act. We are drafting a final rule that responds to the numerous timely comments we received and intend to publish the final rule as expeditiously as possible. Since December 2003, we have been receiving, reviewing, and responding to approximately 167,000 notifications each week about articles of food being imported or offered for import into the U.S.

With the prior notice requirement, specific information mandated by the Bioterrorism Act must be submitted to FDA before the imported food arrives in the U.S. This not only allows FDA's and CBP's electronic screening systems to review and screen the shipments for potential serious threats to health (intentional or otherwise) before food arrives in the U.S., but it also allows FDA staff to review prior notice submissions for those products flagged by the systems as presenting the most significant risk and determine whether the shipment should be held for further investigation.

In addition, FDA has been actively working with the analysts at CBP's National Targeting Center to utilize their Automated Targeting System as a supplementary tool to enhance the Agency's ability to focus attention on those imported foods that may pose a serious threat to public health. This allows FDA to screen products against CBP databases containing sensitive criminal and terrorist-related information. Products identified as potentially "high risk" through FDA's and CBP's screening criteria are targeted and undergo a manual, comprehensive "import security review" by FDA's Prior Notice Center that operates 24 hours a day, 7 days a week, every day of the year. FDA uses defined risk factors to select the products for import security reviews, based on intelligence reports and information about the shipper and/or consignee that indicate a potential risk to the U.S. consumer and the domestic market. Prior Notice import security reviews complement the traditional import field examinations. In FY 2005, FDA conducted intensive prior notice import security reviews on 86,187 imported food shipments.

Administrative Detention

Section 303 of the Bioterrorism Act gives FDA authority to administratively detain any article of food for which the Agency has credible evidence or information that the food presents a threat of serious adverse health consequences or death to humans or animals. This authority was self-executing and provides an added measure to ensure the safety of the nation's food supply. Section 303 also requires FDA to provide by regulation procedures for instituting on an expedited basis certain enforcement actions against perishable foods subject to a detention order. On June 4, 2004, FDA published a final rule to implement this section. The rule also includes procedures for detaining an article of food, expedited procedures for detaining perishable foods, and the process for appealing a detention order.

Maintenance and Inspection of Records for Foods

Section 306 of the Bioterrorism Act authorizes FDA to have access to certain records when the Agency has a reasonable belief that an article of food is adulterated and presents a threat of serious adverse health consequences or death to humans or animals. It authorizes the Secretary to publish regulations to establish requirements regarding the establishment and maintenance, for not longer than two years, of records by persons (excluding farms and restaurants) who manufacture, process, pack, transport, distribute, receive, hold, or import food. On December 9, 2004, FDA published a final rule to implement this section. The recordkeeping regulation requires persons receiving or releasing food, including food ingredients, to identify the immediate previous sources of that food and the immediate subsequent recipients of that food; thus, this rule is often referred to as the "one up/one down" rule. The regulation enhances FDA's ability to track and contain foods that pose a threat of serious adverse health consequences or death to American consumers from accidental or deliberate contamination of food. Affected persons with 500 or more full-time equivalent employees had to be in compliance with the regulation on December 9, 2005. Smaller companies, which provide more than 80% of the food supply, have until June or December 2006 to be in compliance, depending on the number of employees they have. The Bioterrorism Act required FDA to consider the size of the business in developing the regulations. FDA exercised this discretion by giving smaller businesses more time to comply to enable them to learn from the experiences of their larger counterparts and thereby reduce costs.

Authority to Commission Other Federal Officials to Conduct Inspections

Section 314 of the Bioterrorism Act authorizes the Secretary to commission other Federal officers and employees to conduct examinations and investigations. Pursuant to this new authority, FDA and CBP have signed a Memorandum of Understanding to commission CBP officers to conduct examinations and investigations pursuant to information obtained through the prior notice requirements. These examinations and investigations may be carried out on FDA's behalf at ports where FDA may not currently have staff or to augment FDA staff at ports that do have an FDA presence. This unprecedented FDA-CBP collaboration significantly strengthens our ability to secure the border while ensuring the movement of legitimate trade. In accordance with this authority, FDA has already commissioned 9,948 CBP officers. The Agency will continue to explore use of this authority with other agencies with whom we share jurisdiction over a facility as a tool to further improve efficiencies.

INDUSTRY GUIDANCE AND PREVENTIVE MEASURES

In 2003, FDA has issued guidance on the security measures the food industry may take to minimize the risk that food will be subject to tampering or other malicious, criminal, or terrorist actions. FDA issued such guidance, "Security Preventive Measures Guidance Documents," for food producers, processors, and transporters, for importers and filers, for retail food stores and food service establishments, and for cosmetic processors and transporters. In addition, we have issued specific security guidance for the milk industry. During domestic inspections and import examinations, FDA's field personnel, as well as our state counterparts, continue to hand out and discuss these guidance documents to firms that have not previously received them.

To help reduce the risk of an attack on the food supply, FDA and our partners at USDA have joined forces to provide a food security awareness training program entitled, "Protecting the Food Supply from Intentional Adulteration: An Introductory Training Session to Raise Awareness." The training is directed at individuals who play an important role in defending our nation's food from attack: Federal, state, local, and tribal food-industry regulators; school food authorities; and nutrition assistance program operators and administrators. Representatives from the food industry and individuals essential in responding to a food emergency due to an intentional attack – such as law enforcement, public health, and homeland security officials – also are encouraged to participate in the training program. The program is available to any interested individuals free of charge.

VULNERABILITY AND THREAT ASSESSMENTS

FDA has adopted a risk-based approach to address food defense and determine where to focus its resources. As part of our efforts to anticipate threats to the food supply, we have conducted extensive scientific vulnerability assessments of different categories of food, determining the most serious risks of intentional contamination with different biological or chemical agents during various stages of food production and distribution. FDA's initial assessment utilized an analytical framework called Operational Risk Management (ORM) that considers both the severity of the public health impact and the likelihood of such an event taking place. As part of this process, FDA has incorporated threat information received from the intelligence community.

To validate our findings, FDA contracted with the Institute of Food Technologists to conduct an in-depth review of ORM and provide a critique of its application to food security. This review validated FDA's vulnerability assessment and provided additional information on the public health consequences of a range of scenarios involving various products, agents, and processes.

FDA also contracted with Battelle Memorial Institute to conduct a "Food and Cosmetics, Chemical, Biological, and Radiological Threat Assessment." The assessment also affirmed the findings of FDA's ORM assessment. In addition, it provided another decision-making tool for performing risk assessments. Further, the Battelle assessment made a number of recommendations that addressed research needs, the need for enhanced laboratory capability and capacity, and the need for enhanced partnerships between Federal, state, and local governments to ensure food security. FDA is addressing each of these recommendations.

The ORM approach provided a high-level view of foods and agents that were of greater concern. Since the completion of the ORM, FDA has undertaken more in-depth vulnerability assessments of specific food commodities using a method called CARVER+Shock. This method uses processes adapted from techniques developed by DoD for use in assessing the vulnerabilities of military targets to asymmetric threats. Results of these updated assessments are being used to develop technology interventions and countermeasures, identify research needs, and provide guidance to the private sector.

In 2003, FDA began using the CARVER+Shock analytical tool to perform vulnerability assessments to identify what an individual or group, intent on doing damage to the food and agriculture sector, could potentially do based on the person's or group's capability, intent, and past history. The CARVER+Shock methodology was modified under Homeland Security Council leadership for use in the food and agriculture sector by FDA, USDA, and DoD with coordination by DHS, CIA, and FBI. FDA's approach has been to seek voluntary, mutually beneficial partnerships with various segments of the food industry. We have completed such cooperative assessments with segments of the regulated industry that involve bottled water, fluid dairy products, juice products, and infant formula. FDA also has collaborated with USDA to provide assistance to the USDA Food and Nutrition Service on the use of this analytical tool on specific commodities in the school lunch program.

In recent months, FDA has been part of a joint federal initiative along with USDA, DHS, and the FBI called the Strategic Partnership Program in Agroterrorism (SPPA). The SPPA initiative is again using the CARVER+Shock tool but, by seeking industry and state volunteers, is taking the tool to local venues. During these assessments, there are local industry, FBI, DHS, FDA, and USDA participants. These assessments not only identify vulnerabilities in other food commodities but also build local infrastructure around food defense issues. The SPPA program will run for two years and has a goal of completing 40-50 assessments during this period. The results from these assessments will be used to identify mitigation strategies and to focus research questions.

EMERGENCY PREPAREDNESS AND RESPONSE

FDA has established an Office of Crisis Management to coordinate the preparedness and emergency response activities within FDA and with our Federal, state, and local counterparts. Over the past few years, FDA has participated in and conducted multiple emergency response activities including exercises coordinated with other Federal and state agencies. For example, FDA and USDA's FSIS have focused on strengthening our working relationships through joint testing of several response plans in an exercise environment. FDA has participated in numerous exercises, including those sponsored by USDA/APHIS, that focus on the occurrence of natural or intentional outbreaks in animals. We have conducted exercises to test our emergency response with respect to contamination of the food supply and animal feed. FDA also has reviewed food defense and rapid response and recovery procedures with industry groups and trade associations.

To enhance FDA's ability to manage, plan for, and respond to food emergencies, FDA has implemented the Emergency Operations Network Incident Management System (EON IMS), an electronic system for managing emergencies. It has three components: incident tracking and contact management, a collaboration and knowledge management tool for meetings and document management, and a Geographic Information System for mapping and impact assessment. The EON IMS is important in all emergencies and exercises requiring efficient receipt and dissemination of large volumes of information to our stakeholders, including the public and other Federal and state agencies. Once completed, this system will provide a web-based connection for all FDA offices and our partners, through which accurate real-time information about various incidents can be shared and discussed. It will be a component of a safety net that enhances our ability to prepare for a terrorist attack and respond should an attack occur. The development of this system conforms to HSPD-5, "Management of Domestic Incidents."

LABORATORY ENHANCEMENTS

An additional step in enhancing our response capability is to improve our laboratory capacity. A critical component of controlling threats from deliberate food-borne contamination is the ability to rapidly test large numbers of samples of potentially contaminated foods for a broad array of biological, chemical, and radiological agents. To increase surge capacity, FDA has worked in close collaboration with USDA's FSIS to establish the Food Emergency Response Network (FERN) to include a substantial number of laboratories capable of analyzing foods for agents of concern. We are seeking to expand our capacity through agreements with other Federal and state laboratories. There are 123 laboratories representing all 50 states and Puerto Rico that have satisfactorily completed the FERN laboratory Qualification Checklist, which provides vital information to determine if a lab meets the criteria for participation in FERN and is eligible for Federal funding. In FY 2005, FDA was able to offer cooperative agreements to 8 State chemical laboratories which enhanced the current capability and capacity of the 10 FDA laboratories participating in FERN. Participation continues to grow. FERN will encompass a nationwide network of federal, state, and local laboratories working together to build the capacity to test the safety of thousands of food samples, thereby enhancing the nation's ability to swiftly respond to a terrorist attack.

We also are expanding federal, state, and local involvement in our eLEXNET system by increasing the number of laboratories around the country that participate in this electronic data system. eLEXNET is a seamless, integrated, web-based data exchange system for food testing information that allows multiple agencies engaged in food safety activities to compare, communicate, and coordinate findings of laboratory analyses. It enables health officials to assess risks and analyze trends, and it provides the necessary infrastructure for an early-warning system that identifies potentially hazardous foods. At present, there are 113 laboratories representing 50 states and the District of Columbia that are part of the eLEXNET system with 95 actively submitting data. We are continuing to increase the number of participating laboratories and types of data being submitted into the system. Moreover, the governments of Canada, Mexico, and the United States agreed to establish a pilot to use eLEXNET to share food sample data among the three countries' laboratories. FDA has been working with Mexico and Canada to establish a secure network to facilitate the sharing of food-testing data between U.S., Mexican, and Canadian laboratories.

FDA also is collaborating with CDC, USDA, DHS, EPA and many other Federal agencies to create a Memorandum of Agreement for an Integrated Consortium of Laboratory Networks (ICLN). The ICLN will be an integrated system of laboratory networks, such as FERN, CDC's Laboratory Response Network (LRN), and USDA's National Animal Health Laboratory Network (NAHLN), to provide for early detection and effective consequence management of acts of terrorism and other events involving a variety of agents and more than one section or segment of the nation (i.e., humans, animals, plants, food, the environment). The LRN includes approximately 150 domestic and international laboratories. These laboratories are primarily responsible for testing human specimens and a subset of labs can test animal specimens. There are 39 laboratories in USDA's NAHLN that are primarily responsible for testing animal samples.

In addition, FDA collaborated with the U.S. Department of the Army to design and develop two mobile laboratories to be deployed at borders, ports, or other locations, to enhance our ability to provide timely and efficient microbiological and chemical analyses of imported food. FDA took possession of the completed mobile laboratories in April 2005. The Microbiological Mobile Laboratory Unit was utilized in October 2005 during an emergency deployment to assist the Department of Health and Hospitals Laboratories (DHHL) in Louisiana. The DHHL laboratories were unable to provide analytical support due to the damage caused by Hurricanes Katrina and Rita. FDA and the DHHL analysts worked together in the Mobile Laboratory to provide on-site analytical testing of water samples from shellfish-growing waters.

RESEARCH

To prioritize research needs and avoid duplication, FDA coordinates with its sister agencies within HHS, such as CDC, and with other Federal partners such as USDA, DHS, DoD, and the Department of Energy. Within FDA, we have embarked on an ambitious research agenda throughout the Agency to address potential terrorist threats. To increase focus on food defense, FDA is ensuring that significant resources are focused on priority food safety and defense issues. For example, research sponsored by FDA's Center for Food Safety and Applied Nutrition is aimed at developing the tools essential for testing a broad array of food products for a multiple number of biological and chemical agents. We are actively working with our partners in government, industry, and academia to develop such methods. FDA's work with AOAC International, an association of analytical chemists, on validating analytical methods for the detection of biological, chemical, and radiological agents in foods is considered the "gold standard" against which other validations programs are judged. Likewise, FDA's research on microbial genomics and analytical chemistry is widely recognized for its importance to other Federal agencies charged with forensic investigations of terrorism events.

Section 302(d) of the Bioterrorism Act directs FDA to provide for research on tests and sampling methodologies designed to test food to detect adulteration rapidly, particularly methodologies that detect intentional adulteration and tests that are suitable for inspections of food at ports of entry to the United States. This section also requires the Agency to report annually to Congress on its progress. FDA has submitted its second annual report to Congress. It can be found on FDA's Bioterrorism Act webpage (http://www.fda.gov/oc/bioterrorism/bioact.html).

FDA began focusing its research program to address food defense concerns soon after the events of September 11, 2001. The report mentioned above describes more than 100 intramural and extramural research projects to develop tests and sampling methodologies for the detection of adulterated food. The Agency's research agenda is particularly focused on methods to detect high-priority biological agents (e.g., *Clostridium botulinum* neurotoxins) as well as chemical (e.g., ricin), and radiological threat agents that pose the greatest threats to the public and is focused on foods believed to be the most vulnerable or attractive to terrorists. Our researchers also are exploring food-testing protocols using the latest technologies, such as the optical affinity biosensor technology and the quadruple time of flight mass spectrometer, to improve timeliness and accuracy over existing techniques. Researchers are also gleaning information on test methods by using them in studies focused on interventions or shields for the food supply, studies focused on characterizing the behavior (growth, survival, stability) of agents in various food categories, and studies focused on decontaminating food processing facilities.

Among the Agency's research accomplishments are the development, adaptation, or validation of rapid and field-deployable methods to detect various agents in food and the establishment of testing protocols. FDA has shared these new data and technologies with Federal, state, and local entities to equip them to perform food safety testing. FDA has also shared research findings with industry in order to further protect the food supply from deliberate attack.

HHS AVIAN INFLUENZA EFFORTS

Finally, I would be remiss, Mr. Chairman if I did not mention efforts underway at HHS with regard to Avian Influenza. Recent events affecting public health including SARS, Monkeypox and Avian Influenza have highlighted the potential adverse health effects of human interaction with animals. Outbreaks of zoonotic disease are occurring with increasing frequency, from all corners of the world. It is difficult to predict when and where the next event will occur. It is apparent, however, that the public health and agriculture sectors must seek new partnerships and new ways to detect these microbial threats.

Department of Health and Human Services (HHS) Secretary Mike Leavitt has made influenza pandemic planning and preparedness a top priority. The FDA and other agencies within HHS are working together formally through the Influenza Preparedness Task Force that Secretary Leavitt has chartered to prepare the United States for this potential threat to the health of our nation. The Department is also working with other federal, state, local and international organizations to ensure close collaboration.

As you are aware, the potential for a human influenza pandemic is a current public health concern with an immense potential impact. Inter-pandemic (seasonal) influenza causes an average of 36,000 deaths each year in the United States, mostly among the elderly and nearly 200,000 hospitalizations. In contrast, scientists cannot predict the severity and impact of an influenza pandemic, whether from the H5N1 virus currently circulating in Asia and Europe, or the emergence of another influenza virus of pandemic potential. However, modeling studies suggest that, in the absence of any control measures, a "medium-level" pandemic in which 15 percent to 35 percent of the U.S. population develops influenza could result in 89,000 to 207,000 deaths, between 314,000 and 734,000 hospitalizations, 18 to 42 million outpatient visits, and another 20 to 47 million sick people. The associated economic impact in our country alone could range between $71.3 and $166.5 billion. A more severe pandemic, as happened in 1918, could have a much greater impact.

There are several important points to note about an influenza pandemic:

- A pandemic could occur anytime during the year and could last much longer than typical seasonal influenza, with repeated waves of infection that could occur over one or two years.
- The capacity to intervene and prevent or control transmission of the virus once it gains the ability to be transmitted from person to person will be extremely limited.
- Right now, the H5N1 avian influenza strain that is circulating in Asia among birds is considered the leading candidate to cause the next pandemic. However, it is possible that another influenza virus, which could originate anywhere in the world, could cause the next pandemic. Although researchers believe some viruses are more likely than others to cause a pandemic, they cannot predict with certainty the risks from specific viruses. This uncertainty is one of the reasons why we need to maintain year-round laboratory surveillance of influenza viruses that affect humans.

- We often look to history in an effort to understand the impact that a new pandemic might have, and how to intervene most effectively. However, there have been many changes since the last pandemic in 1968, including changes in population and social structures, medical and technological advances, and a significant increase in international travel. Some of these changes have increased our ability to plan for and respond to pandemics, but other changes have made us more vulnerable.
- Because pandemic influenza viruses will emerge in part or wholly from among animal influenza viruses, such as birds, it is critical for human and animal health authorities to closely coordinate activities such as surveillance and to share relevant information as quickly and as transparently as possible.

In the United States, USDA and the Department of the Interior coordinate most work on avian influenza viruses among birds and other animals. HHS collaborates with USDA and the Department of the Interior in critical partnerships for domestic preparedness for a possible avian influenza outbreak in the United States. HHS relies on USDA for domestic and wild bird, backyard bird, live bird market and poultry products surveillance as a way to early detect threats to human health. Early detection will allow the US Government to have the most up-to-date and reliable information that will help to save human lives.

As you are aware, the President requested additional FY 2006 appropriations in support of his National Strategy on Pandemic Influenza. In seeking this funding, the goals are: to be able to produce a course of pandemic influenza vaccine for every American within six months of an outbreak; to provide enough antiviral drugs and other medical supplies to treat 25 percent of the U.S. population; and, to ensure a domestic and international public health capacity to respond to a pandemic influenza outbreak.

CONCLUSION

In conclusion, HHS is making significant progress in its ability to ensure the safety of the food supply. Due to the enhancements being made by FDA and other agencies and due to the close coordination between the Federal and State food safety, public health, law enforcement, and intelligence-gathering agencies, the United States' food safety and defense system is stronger than ever before. Although we are better prepared than ever before, we are continuously working to improve our ability to prevent, detect, and respond to terrorist threats.

Thank you for this opportunity to discuss our counterterrorism activities to protect the food supply. I would be pleased to respond to any questions.

COMMONWEALTH OF PENNSYLVANIA
Edward G. Rendell, Governor

Testimony
Prepared for
The United States Senate
Committee on Agriculture, Nutrition and Forestry
Subcommittee on Research, Nutrition, and General Legislation

Present by:
Dennis C Wolff, Secretary
Pennsylvania Department of Agriculture

January 9, 2006

On behalf of the Pennsylvania Department of Agriculture, and the entire agricultural community, thank you for choosing the 90th Pennsylvania Farm Show as the venue for your Senate Agriculture Subcommittee on Research, Nutrition and General Legislation Field Hearing. The Farm Show is an excellent backdrop to symbolize the diversity and importance of agriculture to our economy, and why it is so important to address the bio-security needs of this industry. As is evident this week, agriculture impacts many people, not just the farming community. And it is this premise that has guided our bio-security planning efforts. I am very pleased that your distinguished Sub-Committee is taking the time to investigate and study how Pennsylvania is successfully implementing bio-security measures to prepare and protect our agricultural system, and American citizens.

Thanks to Governor Rendell, the Pennsylvania Department of Agriculture participated in the 2003 Homeland Security Office of Domestic Preparedness State Homeland Security Assessment and Strategy Program, in cooperation with the Pennsylvania Emergency Management Agency. Pennsylvania was one of six states that included agriculture as one of the disciplines to be evaluated. The Agriculture Assessment reflects the Commonwealth's recognition of their discipline as representative of the largest industry in Pennsylvania and subsequent need to ensure bio-security prevention, response and recovery capabilities. That has guided how we have approached our own programming and partnerships with our federal partners. It has also helped to shape our federal grant requests, aiming to target available funding to specific needs.

Agriculture in Pennsylvania is responsible for approximately $4.5 billion of cash farm receipts and nearly $45 billion in economic activity, making good bio-security planning essential. The economy of Pennsylvania and the nation is very vulnerable to any major disease outbreak due to the closing of export markets. Even a small, rapidly contained outbreak of a disease can cause the loss of millions in export dollars. In 2003, a single case of mad cow disease led to billions of dollars in lost exports for the United States. An extended or widespread outbreak of a disease in the United States would bring on economic disaster.

To protect public health, food supply, and the economy of Pennsylvania and the United States, rapid efficient containment and eradication of diseases or containment of animals, plants and the food supply is vital and has been the driving force behind our preparations at the Department of Agriculture. Laboratory diagnostics capabilities are a vital part of rapid disease identification, which acts as a trigger for disease containment and eradication efforts. The infrastructure for developing response capabilities for agricultural emergencies must be in place before a response to an actual emergency is necessary.

In Pennsylvania, we are very fortunate to have excellent food, animal and plant laboratories that form the basis of our response capabilities. Framed around this basis has been our request for federal assistance that has included updated laboratory information management systems and improved diagnostic capabilities for the Pennsylvania Veterinary Laboratory System; a complete rebuild of our food safety and inspection reporting system; improved diagnostic capabilities for our plant protection systems and continued educational and training initiatives for members of the animal response teams as part of the Pennsylvania Animal Response Team (PASART). This comprehensive approach is in recognition that our nation has only a 3-5 day supply of food on hand at any one time; if an incident involving any part of agriculture is not controlled or contained rapidly and efficiently, we as a nation could face significant human and economic hardship. The Commonwealth has worked cooperatively with the many federal, state and private partners to assess, plan and train for bio-security disasters.

Through my comments today, I will provide an overview of how Governor Rendell's multi-agency approach to bio-security has positioned Pennsylvania as a leader in both planning and preparedness for the operational responses required for accidental or terroristic threats to agriculture. I appreciate that one of the goals of this field hearing is to better understand Pennsylvania's ability to coordinate research and develop an integrated strategy to protect the industry from accidental or deliberate pathogen or biological agent introduction. Let me provide a few examples of what we are doing.

First, in partnership with the Pennsylvania Department of Health and the Centers For Disease Control, we worked to develop a Food Safety and Security System. The new food safety computer system—the Digital Health Department—will enhance the safety and security of the Commonwealth food supply by enhancing communication ability, information analysis and sharing, and uniformity and continuity of those agencies involved in food safety. All Department of Agriculture regulated food industries, from restaurants to food manufacturers, to warehouses will have inspectional, licensing, and contact information housed within one system. And the ability will exist to integrate all local and county health departments into a statewide information and communications network, ultimately providing a statewide database of all regulated food industry and regulatory authorities.

Second, in partnership with PEMA and USDA, we developed a Laboratory Information Management System and an Infrastructure for Identification of Animal Premises Database. The Pennsylvania Department of Agriculture Bureau of Animal Health and Diagnostic Services completed a project to allow consolidation of the Bureau's databases and a link to the Pennsylvania Agriculture Department Laboratory System—LIMS has been completed. This new system allows the Department to share its diagnostic information with state and federal veterinary and medical officials. Also it sets the stage for Pennsylvania's participation in the voluntary national animal identification program.

Third, in partnership with PEMA, we worked on upgrading the plant pathology lab for testing of crop disease organisms. Plant disease-causing microbes could be used to reduce food supplies and to disrupt trade in agricultural commodities that would impact national economics. Historically, plant diseases have caused serious damage to the U.S. economy. Some examples of these diseases include wheat rust, corn leaf blight, plum pox, potato mop top virus, citrus canker and karnal bunt on wheat. USDA has identified the listed several plant diseases and toxins that pose a severe threat to a number of crops.

Fourth, the Pennsylvania State Animal Response Team is a coordinated effort between several governmental, corporate, and private entities dedicated to the preparation, planning, response, and recovery of animal emergencies in Pennsylvania. The team's mission is to develop and implement procedures and train participants to facilitate a safe, environmentally sound and efficient response to animal emergencies on the local, county, state, and federal level. PASART has also played a leadership role at the national level, as its Executive Director, Joel Hersh, helped facilitate a national animal response team dialogue which included representatives from a number of states, as well as representatives of the Department of Homeland Security and USDA, Animal Care Division. PASART and its volunteers at the county level, known as CARTS, as critical to preparedness efforts for our state. The expertise provided, through trained volunteers, will be critical to our mitigating the effects of any disaster, including those caused by terrorism. Funds for SART type programs needs to be included as a line item in any federal preparedness funding allocations.

Fifth, PDA conducted an Avian Influenza discussion-based exercise on December 14, 2005. The exercise participants include personnel from: University of Pennsylvania (PADLS), PSU (PADLS), Bureau of Animal Health Regional and HQ staff, the PA Veterinary Laboratory, PDA EPLO's, PDA EOC personnel, PDA Administrative Services, PA Dept of Health, PEMA, FBI, USDA-APHIS & PennAg Industries. The scenario that PDA exercised began at initial detection of a suspected highly pathogenic AI virus on a fictional large-scale Lancaster County layer poultry operation and concluded once the National Veterinary Services Laboratory (NVSL) determined the AI strain was H5N1 (Once H5N1 is determined the exercise will conclude)

The objectives of the exercise were to: (1) Detect, contain and indemnify the virus utilizing the AI Response Plan; (2) Assess internal and external resource coordination and notification protocols; (3) Implement and utilize the Incident Command Structure and discuss private and public sector coordination; (4) Discuss media control. An After Action Report (AAR) was completed and the following issues were identified as areas to improve or develop:

1. Lack of Personal Protective Equipment (PPE) and training for our responders to adequately protect themselves from an AI that would be dangerous to humans.
2. Improve the notification process in an AI incident.
3. Improve our logistical capabilities (e.g. more contracts, mass depopulation capabilities, manpower resources, agreements with other agencies for additional response efforts & resources).

From the exercise, the evaluation team and members of PEMA recommended PDA explore the possibility of creating trained and equipped response teams that would be fully capable of responding to any type of dangerous animal or emergency disaster that affects agriculture and our emergency response expectations (BPI, Food Safety or Animal Health).

Sixth, Governor Rendell and the Pennsylvania General Assembly have approved the construction of a BSL-3 lab as an annex to the Pennsylvania Veterinary Laboratory at PDA. This additional capacity will allow Pennsylvania to leverage the current investment made by the state for animal health, rapidly respond to the diagnostic needs of our farmers, extend the USDA partnership and more importantly protect our citizenry. This investment will be very beneficial to protecting Pennsylvania's economy.

Finally, the recognition of the possibility that accidental and intentional introductions may occur has emphasized the need to develop rapid response capabilities in order to protect natural and cultivated plant resources from plant pests. For the purpose of this document the term "rapid response" is defined to mean a series of coordinated activities involving one or more organizations that are initiated by the discovery of a plant pest of concern. Rapid response activities consist of a number of components including detection, accurate identification and/or diagnostics, and mitigation activities.

It is unlikely that any single agency or organization has both the legal authority and a sufficient level of resources to conduct an effective plant pest mitigation response without the involvement and support of others. It is important for all involved to clearly understand their roles and responsibilities in a rapid response situation. Generally, a state or federal plant pest regulatory agency will have to be the lead agency in conducting any response activity because of the legal responsibilities assigned to that agency.

Accurate and timely plant pest identification and/or diagnostic support is critical to developing and implementing effective mitigation activities. Therefore, it will be especially important for agencies that have skills in this area be included in the planning and implementation of mitigation activities. Organizations with plant pest identification and/or diagnostic capabilities may be one of the first to receive samples or initial indication of a potential problem. It is essential that this information be provided to regulatory agencies as soon as possible so that effective evaluation and mitigation activities can be initiated.

There are so many other initiatives I could highlight—some small—others large, but all essential to protecting Pennsylvania's health and safety. Everything we have done has been a collaborative effort among private, federal, local and state partners. While we have made great progress, there are several things the federal government could work with us and be helpful on. They include:

1. Support for BSL-3 Lab Annex—Additional federal funding would enable Pennsylvania to position itself as a primary provider of bio-security laboratory capacity for the east coast of the United States.
2. Funding to extend the Garrison system statewide—This would enable Pennsylvania to provide a seamless system for food security to each and every town, hamlet and city in Pennsylvania.
3. PPE equipment and infrastructure—This equipment would finish preparedness planning in the case of an accidental or terroristic incident on a farm.
4. Additional professional staff—Federal funding at the personnel level has enabled Pennsylvania to excel at many of its program and planning levels. Further support would allow us to build upon those successes, and include new areas of bio-security preparedness and planning.
5. Ag Research and Funding—A strong research program fueled by adequate funding is the building block to protecting American homeland security. Research drives many of the successes Pennsylvania has had with our projects and regulatory responsibilities.

Thank you for the opportunity to testify before your Sub-Committee today. I am hopeful my comments will give you a well-rounded view of what Pennsylvania has accomplished and what our future needs are. Thank you and please enjoy your stay at the Farm Show.

Subcommittee on Research, Nutrition, and General Legislation
Senate Committee on Agriculture, Nutrition, and Forestry
Pennsylvania Hearing: January 9, 2006

Testimony By:

Robert D. Steele, Ph.D.
Professor and Dean
Pennsylvania State University
College of Agricultural Sciences
University Park, PA 16802
(814) 865-2541
Rsteele@psu.edu

Senator Santorum, thank you for the opportunity to present testimony to the Subcommittee on Research, Nutrition, and General Legislation today on **Bio-Security: Is U.S. Agriculture Prepared to Protect Itself?** My name is Bob Steele and I am Dean of the College of Agricultural Sciences at Penn State University. I am currently in my 9th year in this position. In addition, I serve in various leadership capacities in the coordinated network of Land Grant University Colleges of Agriculture in partnership with the USDA through the Cooperative State Research, Education and Extension Service (CSREES). As requested, I will limit my remarks to the 7 questions raised and conclude with some general summary comments.

Overview of how Penn State fits within the national structure concerned with agricultural bio-security

Penn State is one of the nation's premier comprehensive research universities and is the Land Grant University for the Commonwealth. Chartered in 1855, it is one of the oldest land grants, awarding the nation's first undergraduate and graduate degrees in agriculture. It is the only university in Pennsylvania offering comprehensive research, education and extension programs across the entire food and fiber spectrum. Our programs reach each of the 67 counties of Pennsylvania and are coordinated with programming in the more than 3,000 counties throughout the United States. This is the essence of the Land Grant University system we've had in place since 1862 and it has been studied and emulated throughout the world. Within the College of Agricultural Sciences, there are approximately 2,500 undergraduate and graduate students pursuing their education with aspirations of becoming some of tomorrow's leaders in these programs in Pennsylvania, throughout the United States and around the globe. Extensive research, education and extension programs centering on agricultural bio-security are coordinated from the college but expertise in this area is located throughout the entire university. The nearly $160M funding portfolio of the college is drawn from federal, state, local and private sources. **All** funds are allocated and re-allocated into priority areas based upon comprehensive and ongoing strategic planning activities involving extensive stakeholder input. . There are no entitlements. Organizations such as those present at this hearing (Penn Ag Industries, PA Farm Bureau, PA Department of Agriculture, other universities, and the Penn State Ag Council to name a few) are important avenues for us to reach stakeholders

Specific initiatives Penn State has taken or services it offers to advance bio-security within the agricultural sector

There are too many ongoing activities in bio-security to mention within the limited constraints of this testimony; however, I will describe five specific activities that document our role as well as how we attempt to coordinate and complement our activities with institutions and agencies within PA as well as with other states.

Animal Health: Avian Influenza. The PA Animal Diagnostic Laboratory System (PADLS) within the Commonwealth has become a model nationally for dealing with animal health issues including both prevention and intervention. Comprised of a partnership among Penn State, the PA Department of Agriculture and the University Of PA School Of Veterinary Medicine, each institution brings its unique expertise to bear on animal health issues important to Pennsylvanians. As others on this and other panels I am sure will testify, the issue of avian influenza (AI) is especially relevant. One lesson learned from past AI outbreaks was the need for better diagnostic tools to more quickly identify viral strains emerging. To develop a better diagnostic requires a more complete understanding of the fundamental biology of the organism, i.e. closing the knowledge gaps, emphasizing the paramount importance of the need for ongoing research, ranging from the very basic to the very applied. The possibility of a highly pathogenic strain of AI arising which can mutate and jump to humans is of course very much in the news today. It should be emphasized that the unintentional (or natural) or the intentional appearance of the virus doesn't matter because **the science is the same**. An avian virologist at Penn State has developed a novel rapid diagnostic test to identify the particularly dangerous H5 and N7 subtypes in only a few hours rather than the many days it takes for traditional virus isolation tests. During a 2001 outbreak of AI in PA, this test enabled a quick diagnosis helping state officials contain the outbreak and limit the losses to 140,000 birds and about $350,000 in direct costs. Contrast this to an outbreak of the same virus a few months later in VA which took much longer to diagnose with a different test. The outbreak in VA resulted in producers losing almost 5 million birds at a cost of over $100 million.

Our researcher has been in Laos and Cambodia with the Food and Agriculture Organization (FAO) of the UN to assist these countries by setting up laboratories to manage avian flu outbreaks there. Working with FAO scientists, our PADLS laboratory system will train them to learn AI diagnostic and surveillance techniques. And finally, scientists at Penn State and elsewhere continue with their research in gaining an even better understanding of basic biological aspects of virology and other animal pathogens so that better prevention and intervention strategies may be developed.

Plum Pox Virus. A few years ago, Pennsylvania experienced an outbreak of Plum Pox virus that seriously impacted the stone fruit industry in PA. With rapid deployment of experts to the field working with the PA Department of Agriculture, we were able to quickly determine the extent of the virus outbreak and contain it through quarantine and removal of infected trees. Using resources largely from PA and levering federal funds and expertise, the containment was successful but costly to the producer. It was a good partnering example of the state dealing with its unique problem in ways that it knew best but levering federal funds to return to PA to assist. Unfortunately, this outbreak demonstrated how poor our fundamental biological knowledge of this and other plant diseases is since the only remedy offered the producer was the complete removal of infected trees. Again, whether a plant pest or pathogen appears unintentionally or intentionally, **the science is the same**. These considerable knowledge gaps absolutely must be closed through more research so as to gain a better understanding of the biology of these organisms so that more effective surveillance, prevention and intervention strategies may be developed.

Chemical Ecology and Bio-Sensor Development. Insects, largely through their antennae, are the undisputed masters at detection of minute quantities of signal molecules in the atmosphere and eliciting rapid and appropriate responses to these signals. A Penn State team of scientists in our Center for Chemical Ecology is hard at work to "learn from nature" in developing at the molecular through the organism level a thorough understanding of the biological, physical and social basis underlying these magnificent systems. With this improved understanding, they are making rapid progress in developing sensors that may be used as molecular sentinels deployed in field locations and elsewhere to detect either unintentional or intentional introduction of chemical/biological agents. It is an exciting example of work at the interface of two pioneering disciplines, biotechnology and nanotechnology, bringing together scientists, and most importantly students (the next generation), from what at first are seemingly unrelated fields to come together to take important inter- and multi-disciplinary approaches. This work also represents a partnership of multiple universities, government agencies (e.g. Defense, NSF, USDA APHIS, etc) and private entities.

Soybean Rust. With the threat of an invasive plant disease such as soybean rust, the USDA developed a national decision support system relying on a computer forecasting service developed by the Penn State College of Agricultural Sciences. The core of this service uses a weather based decision support product that is already beginning to change the way we approach pest and pathogen surveillance, detection and management. The interactive website involves researchers, extension specialists, industry representatives and USDA officials that currently track the development of the soybean crop as well as the disease's progress throughout the United States (www.usda.gov/soybeanrust). This site is likely the largest scale effort of its kind anywhere to date and will in all likelihood become the model that is used for surveillance, detection, and management of the appearance of numerous other pests and pathogens worldwide. It allows state extension specialists to develop recommendations specific for each state since clearly the approach for PA, for example, will likely be quite different than the approach for Mississippi. Again and most importantly, students at the land grant universities are deeply involved as part of their education, research and training to replace us as tomorrow's experts in these critical national interest areas of agricultural bio-security.

The Extension Disaster Education Network (EDEN). As mentioned above, Penn State through its Cooperative Extension activities has a physical and programmatic presence in each of PA's 67 counties that is also linked throughout the more than 3,000 counties nationwide via the Land Grant University Cooperative Extension Service. For decades, emergency preparedness has been a priority area for extension programming. These programs have spanned areas as diverse as natural disasters, animal health, food safety, and emergency readiness to name a few. Within PA, EDEN coordinates with the PA Department of Agriculture, The PA Department of Health, and FEMA and PEMA (federal and state emergency management). Programs developed at Penn State for Ag Emergency Training for first responders have been designated as national programs that are used nationwide. Penn State's Cooperative Extension experts participate on and interact with other federal agencies such as the FBI Agricultural Terrorism Advisory Committee. The national EDEN meeting was hosted by Penn State in 2004. Finally, Penn State has been designated by the CDC as a Center for Public Health Preparedness with emphasis on using zoonotic diseases of significance with wildlife as sentinels.

Areas where additional work would be fruitful for enhanced bio-security

Each of the examples detailed above illustrate the importance of continuing efforts to close knowledge gaps so that we gain better and better understanding of the biological systems we are dealing with. Our ability to develop new diagnostic tools, prevention and intervention strategies becomes quickly limited by these knowledge gaps. Therefore, committing the resources to support the research, education and outreach activities is pretty obvious. In my view, however, the single most important issue is preparation of the workforce needed in this area. The anthrax episode of a few years ago demonstrated so clearly how poorly prepared we were since there were so few people who had a detailed understanding of anthrax. Many retired scientists were contacted for their expertise since there were so few working scientists in this area. Our ability to attract the best and brightest minds to become educated and work in this area is critical because the issue we are discussing today is indeed a generational one. **We will not solve it in a few years!** We will be dealing with it for many, many years to come and the issue will continue to grow in complexity. We cannot outsource creativity and innovation in agricultural bio-security. Our best and brightest young people have options and they will pursue areas in which they see opportunity for them to make a difference to society. We need to be able to demonstrate to these young people that this is a high priority area with opportunity for them to build satisfying and rewarding careers. This is the unique role for America's universities and especially, although not exclusively, the Land Grant Universities. This is for the public good so the public must be willing to make this a priority and find the resources to ensure that it happens.

Obstacles that would impede the work proposed

The easy answer to this question is simply to say that we need more resources. While true, I would like to describe two critical obstacles that I see-one within Pennsylvania and one at the federal level.

Pennsylvania and especially Penn State are woefully behind the curve in having adequate facilities available of the biosafety level 3 type to conduct the kind of work that is and increasingly will be required for agricultural bio-security of all types. Precious few federal dollars have returned to PA to help and precious few state dollars have been identified. Our three major research universities, Penn, Pitt and Penn State, have worldwide reputations for our cutting edge research in virtually all areas yet we are woefully equipped in this area. Something statewide with federal help needs to be done to address this critical issue. Admittedly, it will be expensive; however, it must be done because the alternative risk is far, far more expensive.

At the federal level, a much more substantial structural issue is apparent. In 1994, the research and extension arm of USDA began to reorganize toward a much more flexible and efficient structure. For many reasons, the re-structuring stalled and it needs to be reenergized. This part of the agency is fraught with inefficiency, redundancy and poor coordination such that the 100 plus year federal-state land grant partnership is at risk. A system that has served worldwide food security so well throughout the 20th century is not prepared to deal with 21st century issues. We need to find ways to better coordinate an already multibillion dollar portfolio within the ARS, ERS, and CSREES as well as in other agencies. I serve on a national task force that has been attempting to re-energize this reorganization. Without question, the obstacles are great but our resolve must be even greater because without it, the outcome will be something nobody wants. This Senate Subcommittee will play a key role in helping to rewrite the authorizing language of the Farm Bill that will be necessary for us, **working together for the common good**, to get it done.

Time and resources needed for implementation of programs

Without question, the post World War II model leading to the development of the National Institutes of Health has been an unqualified success story. In my view, a major reason has been the recognition from early on that large complex research and education projects need stability of funding. What emerged was a more or less 5 to 10 year pattern of funding that could be renewed for additional 5 to 10 year increments based upon past performance and proposed new work addressing priority areas. In short there were no entitlements, yet many projects enjoyed decades long support because they were high quality and of high priority. We need to move closer to this model in agricultural bio-security, not farther from it. Again, this will require re-thinking our organizational structure, particularly with respect to balancing intramural and extramural research and outreach programs. In the end, however, we must be certain that there are adequate resources to support the existing workforce and to recruit and prepare the next generation workforce for the food and fiber sector, particularly in the area of agricultural bio-security.

Identify redundancies where better coordination among parties would increase bio-security

Several federal agencies support work in bio-security ranging from very basic to applied translational work. The new Department of Homeland Security was, in part, set up to help coordinate much of this work. Clearly that has not yet happened to virtually anyone's satisfaction. As one example, the Cooperative Extension funded EDEN network described above is every bit as much of a priority for homeland security and defense as it is for agriculture programs. Wouldn't it make sense for Homeland Security and Defense funding to help underpin and broaden this already in place and time tested successful county based program network than develop yet another stand alone network?

Redundancy is a difficult issue. From a distance, what may appear as a redundancy actually is not upon closer inspection. The issue here is to always question whether what we are doing either is being done somewhere else, or perhaps more to the point, could be done better elsewhere. The key is communication and coordination and to have open assessment and accountability measures in place. I believe that we strive to do this at Penn State.

Specify which issues are within the purview of the government and which devolve to the private sector

Much of this has been addressed above but I would add there can be no question that matters relating to the protection and advancement of our food, water and fiber supply are in the public good. As we've become increasingly aware of late, these are matters in the public interest but also in the interest of our national defense. Therefore, the public through the legislative and appropriations process must be prepared to support these interests through their support of research, education and outreach programs coordinated at the federal, state and local level. Again, this is particularly important with respect to support of the current workforce but also in the identification and education of the next generation workforce. The private sector has an obligation here as well for it is in this sector where many of these individuals will come to be employed. These young people are the future creators of new knowledge and innovators and they will be crucial to the future success of the private sector.

So I conclude by coming back to the question- **Are we prepared?** My answer is- **Not nearly well enough.** Through our existing fund of knowledge we can do a better job and with ongoing support for continuing to close our knowledge gaps in this generation and beyond, we will do a better job. If we don't, then who will?

Thank you again for the invitation to participate and provide testimony on this topic important to all Pennsylvanians and all Americans. I would be pleased to elaborate on anything I've mentioned or to provide any additional information to you, other members of your Committee, or your staff.

Pennsylvania Field Hearing of the
Senate Committee on Agriculture, Nutrition, and Forestry
Subcommittee on Research, Nutrition, and General Legislation
Pennsylvania Farm Show, Harrisburg, PA, January 9, 2006.

Written testimony of
Dr Gary Smith, Professor of Population Biology and Epidemiology, University of Pennsylvania School of Veterinary Medicine.

How the School of veterinary Medicine at Penn fits within the national structure concerned with agricultural biosecurity

The School of Veterinary Medicine at the University of Pennsylvania is one of 27 schools of veterinary medicine in the United States. Our primary contribution to the national effort to maintain agricultural biosecurity involves

- training the veterinarians who will comprise the primary responders to any threat to agricultural biosecurity
- and the research and service component of our work that is explicitly directed towards biosecurity issues (specific examples are listed below)

Our urban clinic, in the heart of the University City section of Philadelphia, draws an enormous caseload of varied small species. And our pastoral New Bolton Center, in Kennett Square, Chester County, 32 miles southwest of Philadelphia, on the edge of Lancaster County, attracts a large number of assorted large animals. During the 1998-1999 academic year, we saw 24,000 small animals and 6,000 large animals. The core/elective curriculum allows students limitless choices. Seniors can select one of five distinct clinical areas: small animal, mixed (small and large) animal, large animal, equine, and food animal medicine. The huge biomedical research complex that comprises the University of Pennsylvania greatly enhances the teaching that students receive and their opportunity to participate in cutting-edge research in countless disciplines.

The School of Veterinary Medicine at the University of Pennsylvania graduates between **110-130 veterinarians** each year. Not all of them will go into to food animal practice, but all of them have received extensive training in zoonotic infectious diseases and in the recognition and treatment of those infectious diseases that may compromise the safety of our food supply or imperil the productivity of our herds and flocks. Up to a third of any graduating class will also have been trained in the construction and use of mathematical models that inform policy decisions concerning the area-wide control of epidemic disease in domestic animals.

Existing services offered by the School of Veterinary Medicine that advance biosecurity within the agricultural sector

- **The Salmonella Reference Center** has become a valuable resource, not only for Pennsylvania farmers, but for other agricultural laboratories in the nation. The

Center is part of the Pennsylvania Animal Diagnostic Laboratory System (PADLS); funding is received from the Pennsylvania Department of Agriculture, along with grants from USDA, the American Egg Board, and with revenue generated through client services.

- **The William B. Boucher Field Service** provides routine and emergency health care for **local dairy,** horse, **llama/alpaca and small ruminant clients.** The dairy and small ruminant specialists offer consultation in production medicine as well as routine on-farm services

- Pennsylvania is the nation's third largest chicken producer, primarily egg-layers, and New Bolton Center is the diagnostic center of choice for many poultry farmers in the region.

- The **Commonwealth of Pennsylvania's Animal Diagnostic Laboratory System** (PADLS), **a third of which is housed at New Bolton Center**, brings scores of cases annually for diagnosis. The laboratories of large animal pathology and toxicology, avian medicine and pathology, microbiology, and clinical laboratory or part of this tripartite laboratory system. Specimens are accepted 24 hours a day. The facilities are credited by the American Association of veterinary laboratory diagnosticians and offer state-of-the-art diagnostic services. A field studies unit at NBC is part of the PADLS. The field studies unit provides the academic and technical support to clinically investigate economically significant animal health and production problems in the Commonwealth of Pennsylvania. Information will be shared with other participants say the programmatic health-care interventions can be developed on a statewide basis.

- The **Center for Animal Health and Productivity** focuses on improving the health of herds and flocks, not just individual animals. This food-animal program looks at the entire farm operation while considering economic, environmental and food-safety perspectives.

Emerging services and initiatives

1. Avian Influenza

(a) The importance of Geographical Information systems: Working with the Pennsylvania Department of Agri-culture and the poultry industry, the School of Veterinary Medicine at the University of Pennsylvania has been instrumental in developing and implementing a statewide avian flu surveillance program after an outbreak of highly pathogenic avian flu among Pennsylvania poultry in 1983–84 that resulted in the destruction of more than 17 million birds at a cost of nearly $65 million. We have a comprehensive strategy that includes constant education and reminders about biosecurity practices, monthly surveillance testing of blood or eggs, regular reviews and updates to our response plans, and ensuring that we have the most advanced technology in our arsenal," she explains. "This includes Geographic Information System (GIS) technology, which is one of the most powerful weapons we have to control diseases and minimize economic loss. Pennsylvania was one of the first in the poultry industry to develop GIS. Since 1998, the School has been using GIS technology to map the location of commercial poultry flocks, feed mills, processing plants, rendering plants, hatcheries, and components of the live-bird market system throughout the Commonwealth. GIS also can be used to create buffer zones around infected flocks for increasing surveillance testing or possible quarantine purposes. The School has proven the value of this state-of-the-art technology. In 1997, just before we began using GIS, there was an outbreak of avian flu in Pennsylvania. Despite a rapid diagnosis of the problem, it took several months to contain the outbreak and cost the Commonwealth of Pennsylvania $3.5 million. In 2001, we experienced another outbreak: with GIS in place and fully functional, we were able to quickly identify where the flock was, where surrounding flocks were, which ones we had to monitor, and which routes trucks should use to avoid infected flocks, among many other applications. GIS gave us the ability to respond very quickly and make very quick decisions. As a result, we had the outbreak under control within one month at a cost of only $400,000, nearly 90 percent less than the cost of the 1997 outbreak.

(b) The importance of rapid diagnosis: The School's ability to respond rapidly to avian influenza was dramatically enhanced in December 2004 with certification by the National Veterinary Services Laboratory to use a new real-time preliminary chain reaction (PCR) test based on technologies similar to human DNA testing used in criminal cases. The real-time PCR takes just a few hours to complete in comparison to tests used previously that take several days. This rapid test enables us to get on top of the situation much faster and, as a result, decrease the cost of outbreaks.

(c) The importance of mathematical modeling to inform policy decisions concerning the control of avian influenza: In December 2005, the School of Veterinary Medicine received almost a million dollars in federal funding to continue our work on mathematical models on avian influenza in poultry flocks. These models are used to explore and rank in order of effectiveness putative strategies for defending against the natural or deliberate introduction of avian influenza virus into poultry flocks in Pennsylvania and adjoining states. As a result of this award, the research group in the Section of Epidemiology and Public Health has become a member of the National Institutes of Health **"Models of Infectious Disease Agent Study"** (MIDAS) and, as such collaborates in this effort with research groups in the USA (at the Pennsylvania State University, for example) and abroad (the University of Warwick, UK). **MIDAS** is a collaboration of research and informatics groups to develop computational models of the interactions between infectious agents and their hosts, disease spread, prediction systems, and response strategies. The models will be useful to policymakers, public health workers, and other researchers who want to better understand and respond to emerging infectious diseases. If a disease outbreak occurs, the MIDAS network will be called upon to develop specific models to aid public officials in their decision-making processes.

2. Surveillance

Surveillance is the bedrock upon which rest all of our strategies to maintain agricultural biosecurity. The **Primary Animal Health Care Specialist Training Program** is designed to train those actually working on dairy farms to recognize diseases and conditions that may affect productivity or portend some serious breakdown of biosecurity. This new program is funded by the private sector (Pfizer) and is a collaborative effort between **the Center for Animal Health and Productivity** at the School of Veterinary Medicine and Delaware Valley College. The idea is to increase the pool of well-trained first responders and to buttress that training with substantial internet resources for diagnosis and the exchange of information.

Areas where additional work is needed and obstacles to progress.

(a) The importance of spatial heterogeneity and the particular problem of applying spatial models in the USA: Mathematical models of infectious disease transmission dynamics inform policy decisions about disease control. The models for SARS and BSE (Mad Cow Disease) transmission dynamics provide two recent examples. A decade ago, most infectious disease transmission models were deterministic, mean-field models that took little account of spatial and stochastic events. Despite the success of these deterministic, mean field models it became apparent that there were situations in which spatial heterogeneities in transmission were important prognostic factors with respect to control - and that these heterogenieties required explicit representation. A burgeoning literature setting out the ways in which such problems could be dealt with in a modeling context and the need to address urgent contemporary problems that clearly involved spatial components has led to the development of a large number of spatial, stochastic models. Examples of these include the several models of Foot and Mouth Disease that were used to inform and examine policy during and after the 2001 FMD outbreak in Great Britain and models of Foot and Mouth Disease in the US. These spatial, stochastic models have been spectacularly successful. Not only do they provide a good representation of an observed outbreak, but they provided an ideal vehicle for comparing and contrasting putative control strategies. What all of these models had in common (both in the UK and in the US) was that **they absolutely depend upon accurate maps of farm locations** and good information concerning the size and type of each farm. In Britain, such information was available on a national scale: agricultural Census included the easting and northing coordinate of the farm-house together with the numbers of pigs, cattle, sheep, goats and deer present on each farm at the data of the census.

There are two important facts here: first, the easting and northing coordinates which constituted the unique farm identifier for census purposes also *provided a good measure of each farm's location*, and, second, *the census data for each farm were available to the modelers*. **Similar information is not in the public domain in the United States, nor (in the opinion of the writer) is it ever likely to be. Issues of confidentiality confound and impede all efforts to gather *and to use* such information in the United States.** The arguments regarding individual and corporate rights to privacy are quite proper but an unintended consequence of these arguments is to deprive the nation of a proven means of defense against the consequences of catastrophic epidemic disease. There very few states in the US where good maps exist for more than just a few. Thus, the US is at a considerable defensive disadvantage.

(b) The importance of a national animal ID system

On 30 Dec 2003, the USDA announced additional safeguards to further minimize the risk for human exposure to BSE in the United States. Among these it was stated that " To enhance the speed and accuracy of the response to animal health threats such as BSE, APHIS is working to implement a national identification system to track animals of various species through the livestock marketing chain." The projected time lines for the establishment of a National Animal Identification System (NAIS) in the draft documents issued by USDA/APHIS (May 2005) propose requiring stakeholders to identify premises and animals according to NAIS standards by January 2008. Requiring full recording of defined animal movements is proposed by January 2009. These same documents acknowledge the very considerable concern voiced by stakeholders and it is clear that many producers are fearful about what might happen should a NAIS become mandatory (there are some voluntary programs currently). For example in March, 2004, the California Cattleman's Association stated that they "will continue to work closely with NCBA and CDFA to ensure that the implementation of any national animal identification system does **not impose unreasonable costs** to producers **and limits the liability of individual producers** and others in the production chain. Of particular significance is determining **how best to protect the confidentiality of producer records**".

"Trace Back" and Trace Forward" methodologies are crucial components of any attempt to curtail and investigate epidemic disease. An effective NAIS would dramatically enhance our nations ability to deal with natural and deliberate introductions, but this writer expects that the legal challenges to such a system will be such that the time lines proposed by USDA/APHIS cannot possibly be satisfied.

Good maps and an effective animal identification system will probably not be available in the short term, so what can we do instead?

(a) Maps and models

If we cannot use existing spatial models (which depend upon accurate mapping of the location of individual farms) can we use spatial models that operate at lower resolutions and still answer worthwhile questions about disease control? This is the problem of granularity. One possible solution might be to turn to the public domain data made available by the US Census. Using census data, it may be possible to map, by zip code, the number of farms of a given type. A number of research groups (include the modeling group in the School of Veterinary Medicine at Penn) have proposed to investigate the utility of metapopulation models as a replacement for detailed spatial models in the design of disease control strategies. The essential insight of metapopulation modeling is that populations exist in "patches" and that we cannot predict the dynamics of the metapopulation as a simple function of the local dynamics within patches because the larger scale processes of migration and colonization, as well as the regional distribution of patches, determine metapopulation dynamics. The models take explicit account of the spatial relationship of the patches and the number of susceptible units in each patch but they do not require any information about the location of each unit within the patch. Thus if we were to define patches in terms of zip codes and infectious units in terms of farms within zip codes, we do not need to know the exact location of each farm. Given that metapopulation models have provided useful insights into observed epidemic patterns in human populations it seems reasonable to hope that the zip-code information contained within the Census of Agriculture could provide the basis for similar metapopulation models of infectious diseases in farmed animals in the US. The funding for this work should reasonably come from state and federal government.

(b) Maps, Animal ID systems and "Trace back": One problem with maps is that they have to be maintained. Another is that it is very difficult to construct maps of farm enterprises from the supplied address because the address often does not correspond to where the animals are actually located. Farms also go out of business or change the nature of their business. Some of these difficulties might be overcome by **"Real Time Mapping"** in which patterns of commerce are recorded using geographical positioning instruments permanently fixed in bulk milk tanks (for example). The patterns recorded by such a system could be used as a surrogate for maps of farm locations and have the distinct advantage that they reflect actual current farming practice (and the amount of milk collected used as an index of herd size). Such a system would require that private sector trucking companies and feed mills (for example) collaborate with information integration centers located in universities or state agriculture departments. Although the examples used here deal with dairy farms, it is not difficult to imagine that the same system could be used to monitor the swine or poultry industries too. Funding would reasonably be the responsibility of state governments.

Respectfully submitted

Gary Smith, January 5th, 2006.

Testimony to the
Pennsylvania Field Hearing
of the
Senate Committee on Agriculture, Nutrition and Forestry
Subcommittee on Research, Nutrition and General Legislation
Pennsylvania Farm Show
Harrisburg, PA
January 9, 2006

Bernard D. Goldstein, MD
Professor of Environmental and Occupational Health
Graduate School of Public Health
University of Pittsburgh

Senator Santorum; Members of the Committee

I am Bernard Goldstein, MD, until recently dean of the University of Pittsburgh, Graduate School of Public Health. My background is that of a physician, toxicologist and public health expert. My past experience includes service as Assistant Administrator for Research and Development at the US Environmental Protection Agency

Since ancient times warfare has included attacks on the agricultural base of an enemy - such as Scipio's ploughing the fields of Carthage with salt. The prudent nation has considered how best to defend this agricultural base, and what investment is appropriate in developing countermeasures. We in public health have been particularly concerned with food safety and with the public health infrastructure, to which I will confine my remarks. The many related important areas under consideration by your subcommitee will I am sure be ably covered by my colleagues from Penn State and the University of Pennsylvania

The University of Pittsburgh Graduate School of Public Health is the only fully accredited school of public health in Pennsylvania, although our colleagues in Drexel are nearing accreditation of another Commonwealth school of public health which we welcome. Our school of public health excels in research, a subject of today's hearing. We are first in the nation among state-related schools of public health in competitive grants from the National Institutes of Health, and only behind Johns Hopkins and Harvard among the 37 schools of public health nationwide. In terms of the Commonwealth, NIH funding to the Pitt Graduate School of Public Health exceeds that of three of the six medical schools, demonstrating the particular vibrancy of public health education and research in Pennsylvania.. The University of Pittsburgh has continued to dramatically increase its NIH funding and its funding from other federal sources, ranking in the top ten nationally for each of its academic health center components and for the university as a whole.
Let me begin by addressing the value of academia to the nation's defense against terrorism. We in the university community recognize that there is an urgency to the shoring up of protections for our nation's food supply. But we also recognize that there is a need for better types of defenses - for new approaches that will increase the effectiveness, as well as lower the costs, of defending the nation's agriculture. There are three major reasons for this.

First, as is evident from the hearing today, our current technical approaches warrant improvement.

Second, the threat to our nation will extend over periods of years, decades and perhaps even generations. We know from history that, no matter how we improve our defenses now, we can not depend upon the defenses of today to be sufficient for the threats of the future

Third, the threat is directed by other humans, not by natural forces. These other humans, our enemy, will respond to any defenses we erect by purposefully attempting to bypass these threats.

The response to the threat to our nation's agriculture must contain a vibrant research activity including our nation's research universities. We have the technical skills to be responsive now, and the track record demonstrating that we can anticipate new threats and respond to the unexpected.

Response of the University of Pittsburgh

The University of Pittsburgh has taken very seriously its responsibilities to respond to threats of biological and chemical terrorism. In addition to specific programs described below pertinent to the topic of today's hearing, there is a campus-wide coordinating program reporting to the Provost - the Center for National Preparedness headed by Dr Kenneth Sochats.

Rural Public Health

Safeguarding American agriculture requires careful attention to the rural infrastructure involved in response to terrorist threats. Along with our colleagues at Penn State, we have major programs focusing on health in rural parts of our Commonwealth. The Graduate School of Public Health has worked closely with the University of Pittsburgh at Bradford on whose campus is a nationally known Center for Rural Public Health Practice, headed by Michael Meit, including being actively involved in the founding and funding of this Center. Response to bioterrorism has been a central part of this program. Of particular note is a major national meeting on the topic of the Rural Public Health Research Agenda organized in 2004 by the Center for Rural Health Practice with HRSA funding. In the nationally distributed document resulting from this meeting, Michael Meit, the head of the Center for Rural Public Health Practice wrote the section on "Rural Public Health Preparedness". The second of eight bullets in this section states:

> "Rural areas are the locus of agricultural production which could be key in targeting the food supply, as well as accessing agricultural chemicals"

The Pitt-Bradford Center participates in programs developed through funding received from the Centers for Disease Control by the Graduate School of Public Health's Center for Public Health Preparedness, headed by Margaret Potter. We are one of over twenty such centers nationally, but one of only five that have significant preparedness activities aimed specifically at rural public health preparedness. Our center was recently awarded $5.4 million for the next five years beginning in fiscal 2005 to train front-line public health workers and first responders. The University of Pittsburgh Graduate School of Public Health's Center for Public Health Preparedness offers onsite and long distance training in preparedness nursing, bioterrorism, preparedness law, forensic epidemiology, and environmental health to a combined total of more than 700 public health professionals and clinicians the year. However, in collaboration with the Center for Rural Public Health Practice at Pitt-Bradford, we could and should do more specifically aimed at rural public health and agricultural issues.

Our task is quite challenging for two reasons, only one of which is insufficient funding. The other is the lack of public health infrastructure in our Commonwealth's rural areas. According to federal HRSA figures, Pennsylvania ranks last among all states in the size of its public health workforce per capita - we are number fifty. As we have significant public health departments in ten of our larger counties and cities, this shortfall is particularly notable in our rural areas. As just one example, many of our Northern tier counties have an infrastructure in place of perhaps a total of 2 public health nurses, while demographically similar neighboring Southern tier counties of New York State have a public health workforce that counts in the dozens. Let me emphasize that we have dedicated and effective public health professionals in rural Pennsylvania, as well as in our state Department of Health. We simply do not have enough of them to mount effective programs, let alone be responsive to new threats.

Our academic program has not waited for the Commonwealth to provide this infrastructure. Rather, we have used our CDC funding to reach out to the entire responder community: firefighters, police, EMT personnel and others active throughout the rural parts of our state. Some of this effort is being performed in collaboration with our colleagues at Penn State. The programs we have developed are beginning to be copied nationally through the interaction of the CDC-sponsored Centers for Public Health Practice and the good offices of the Association of

Schools of Public Health. But much more can and should be done. Supplemental funding specifically aimed at using the expertise of the Center for Public Health Preparedness and the Center for Rural Public Health Practice, at a total of $300,000 per year for five years, would allow penetration of training programs related to agricultural bioterrorism and food safety to all of the rural counties of Pennsylvania, and national dissemination of the innovative training programs and lessons learned.

Biomedical research related to the security of agriculture

We are living in an era of unparalleled advances in biomedical sciences. These advances provide the basis for threats to all life, including our agricultural resources. But they also provide the basis for effective defenses. As just one example, the molecular epidemiology laboratory of Dr Lee Harrison of the University of Pittsburgh has been very active in using new molecular methods for outbreak detection and forensic analysis of bacteria - methods that are more objective, portable, and have much higher throughput than pulsed-field gel electrophoresis, the most commonly used method by public health institutions. Unfortunately, funding to apply this work to terrorist threats has been less than robust. With an additional $250,000 per year for three years, Dr Harrison's laboratory efforts could be readily extended to current and as yet unknown threats to the nation's agriculture, including approaches that would lead to the rapid development of simplified detection methods for any new, unanticipated threat.

There are numerous other examples of the potential for application of cutting edge NIH-funded biomedical research applications to agricultural threats. A multi-million dollar yearly initiative aimed at coupling public health and agricultural research methodology to address biological threats to agriculture would be very much in the nation's interest.

Integrated Assessment and Planning

The Center for Biosecurity of the University of Pittsburgh Medical Center is perhaps the first and certainly among the foremost programs in the US and elsewhere aimed at understanding the broad issue related to the threat of bioterrorism to the nation's health, economic and socio-political security. Its principals are DA Henderson and Tara O'Toole. Dr Henderson is the physician most responsible for the eradication of smallpox; the former dean of the Johns Hopkins School of Public Health; the founder of the first comprehensive institute related to biosecurity issues, and the person to whom President Bush turned to in response to the threat of bioterrorism after Sept 11, 2001. Dr O'Toole, the CEO of the UPMC Center for Biosecurity, is a physician with excellent credentials, including service as Asst Secretary of the Department of Energy responsible for environmental health. The Center for Biosecurity is capable of providing an integrated assessment of the threats to agricultural security that will assess current and future vulnerabilities in relation to broad national objectives. Of particular pertinence is their expertise in understanding the future biological capabilities of terrorist organizations, and the appropriate approach to defense against these capabilities.

Thank you for the opportunity to testify on this important subject. I would welcome answering you questions.

Pennsylvania Field Hearing
of the
Senate Committee on Agriculture, Nutrition and Forestry
Subcommittee on Research, Nutrition and General Legislation
Pennsylvania Farm Show
Harrisburg, PA
January 9, 2006

Testimony of
James L. Adams
Wenger's Feed Mill, Inc.
P.O. Box 26
Rheems, PA 17570
717-367-1195
jadams@wengerfeeds.com

Senator Santorum, Secretary Connor, distinguished members of the Committee, my name is Jim Adams and I am the President and COO of Wenger Feeds, an agribusiness based in western Lancaster County serving the Mid Atlantic region. Wenger's is a family owned and operated company that has grown to become one of the most comprehensive Agricultural Feed and Service companies on the East Coast. Our company is one of the only businesses in North America with three ISO certifications, a recognized international standard of excellence. We are certified in ISO 9001, 14001 and OHSAS 18001 for quality, safety, and the environment as a reflection of our commitment to continual improvement.

Biosecurity is a top priority for our company because it helps protect our customers from diseases which have the potential to cause significant economic loss to their business and the family farmers they work with. Wenger Feeds also owns and manages our own laying hens and has a vested interest in protecting and maintaining the health of our birds.

In the early 1980's, we experienced one of the most catastrophic events that agriculture had ever faced – the outbreak of highly pathogenic avian influenza. The poultry industry, in cooperation with the PA Department of Agriculture and our partner universities, has learned a great deal about biosecurity since then and we have been able to consistently handle these situations with more efficiency and less loss to producers.

In fact, the industry was credited with jump-starting the federal response by organizing bird owners, creating production location maps, and depopulating some of the earliest flocks at the owner's expense. We learned the hard and expensive lesson that timing is critical for controlling disease and missed days increase the risk of spread exponentially.

While our progress has been substantial, there are four areas though where I believe improvement could be made to provide a higher level of security. These areas are detection, indemnity, vaccination, and protection from anti-agriculture organizations.

The first area which has the potential for the greatest impact on biosecurity is detection of an outbreak. Whether the disease is introduced accidentally or intentionally, a rapid diagnosis is critical for controlling an outbreak. The longer a disease has the opportunity to smolder undetected, the higher the probability becomes of additional farms becoming infected or worse that the disease will change into a more virulent form.

We need to have faster, more accurate results to determine the status of individual farms in order to control and eradicate threats. In the 1980's we could see where a few days of hesitation could allow the disease to spread to many of the neighboring farms.

Ideally, there would be a simple and reliable quick test that could be conducted and verified on site. The federal government should be investing in research at land grant institutions and other universities who have tremendous academic knowledge as well as field experience to create the next generation of surveillance tests.

The second area for improvement is in indemnity for producers who participate in controlling and eradicating an outbreak. There is the potential for companies or producers to elect marketing of suspected flocks or animals rather than reporting them because of the potential for significant economic loss due to a limited or slow indemnity program. A stronger, faster, and broader indemnity system would encourage submission and surveillance out in the field and help identify a potential outbreak before it happens.

A complement to a stronger indemnity program would be a federally subsidized business interruption insurance program specifically targeted to producers and family farmers for coverage in the event of a disease outbreak. This could be similar to other federal risk management programs where producers have a minimal cost for insurance to keep their farms operational during an outbreak.

A third area I would like to mention is the use of vaccination during a disease outbreak. Science has shown that vaccination is a very effective method for controlling and eradicating avian influenza and yet international markets tie our hands and do not allow us to use this important tool. There has been some progress with countries recognizing the validity of regionalization for export restrictions but we need to rely on science rather than politics in dealing with disease outbreaks.

Rapid controlled application is critical for vaccine to be used as an eradication tool but production of the necessary materials can take months to prepare. It would be prudent for the federal government to stockpile targeted strains of vaccine to be used in an emergency.

The final area I would like to mention to help guard animal agriculture from devastating diseases is protection from anti-agriculture organizations which have become much more aggressive in their rhetoric and tactics. Animal right groups such as Hugs for Puppies have recently invaded farms here in Pennsylvania with complete disregard to the biosecurity we work very hard to maintain on our farms. Stiffer penalties for domestic terrorist organizations and acts should be enacted to properly reflect the seriousness of breeches in biosecurity and become a deterrent for their activities.

Biosecurity is a very relevant issue for today's agriculture and I commend you for convening this hearing to discuss concerns on how we can build even stronger protections for our industry and family farmers.

Thank you for the opportunity to testify before you today and I look forward to answering any questions you would have.

Testimony
Of
Walt Peechatka, Executive Vice President
PennAg Industries Association
Northwood Office Center
2215 Forest Hills Drive, Suite 39
Harrisburg, PA 17112-1099
717-651-5920 (Phone)
717-651-5926 (Fax)
Before
Senate Agriculture Subcommittee
Pennsylvania Farm Show building
January 9, 2006

PennAg Industries Association appreciates the invitation and the opportunity to provide comments to this panel on the important subject of agriculture biosecurity. PennAg represents more than 600 companies that provide the inputs and services to the more than 50,000 farms in this state.

Pennsylvania is a leader nationally in the area of biosecurity on our poultry farms. After a devastating avian influenza outbreak in the early 1980's, Pennsylvania's poultry industry began an aggressive effort to prevent future outbreaks and minimize their impact if they did occur.

All commercial poultry operations in the state have been included in a data base that can be utilized by the industry, and by our regulatory partners in the USDA and the PDA, in the event of a disease outbreak. This data base has been in use now for more than a decade and has been very effective in identifying neighboring flocks to a poultry flock that has been diagnosed with AI.

It is important to note that the database is maintained by one of our partners at an academic institution thereby enabling us to maintain confidentiality of this information, which is an area of major concern to farmers.

In addition to the excellent database that we have available, the second component of the Pennsylvania program is the surveillance program practiced by our poultry industry. Each year poultry producers send over 250,000 routine samples to our laboratories for diagnosis. This routine surveillance program has enabled us to identify a disease outbreak at its earliest stages and allows us to quarantine, alert neighboring poultry operations, elevate biosecurity operations, and depopulate a flock to control the disease when that is the recommended course of action.

The Pennsylvania program is nationally recognized and several national organizations, specifically the National Chicken Council and the United Egg Producers, have asked for information about our program so they can assist other states in developing similar programs.

We would encourage the US Congress to consider the following items in addressing the issue of biosecurity and developing other initiatives which provide financial support to the agricultural community.

1) Increase the diagnostic laboratory capability in states with large concentrations of poultry and livestock. We encourage the federal government to provide additional financial support for the development of a BSL-3 Laboratory at one of our Pennsylvania research institutions so that the facility can be used for both research and for routine sampling and diagnosis. States with these large concentrations of animals must be self sufficient when it comes to rapid diagnosis. They can not rely on neighboring states or the federal laboratory system if diagnosis is to be prompt and the response rapid.

2) We believe every state should be encouraged to develop a premise identification program along the lines of the program Pennsylvania has developed for the poultry industry. However, the issue of information confidentiality must be addressed when undertaking this effort. Only if the information is confidential will government be able to develop and maintain a database that is accurate. Industry cooperation will be difficult, if not impossible, to obtain if confidentiality is not assured.

3) Agroterrorism has been occurring with increasing frequency. Animal rights groups have been breaking into research facilities and commercial poultry operations for the purpose of releasing animals and filming the contents of the facilities. These activities are strict violations of the basic tenets of biosecurity and place the owners of these operations in jeopardy of having disease introduced to their operations.

   As an example, one of our PennAg members experienced agro-terrorism in the fall of 2005. In this incident, uninvited visitors broke into three different poultry facilities late at night and filmed the inside of these poultry houses. In doing so, they violated every tenet of strict biosecurity.

   To address this problem we recommend that the Federal statutes be strengthened. Federal laws should be such that violations are recognized as a national threat to our country and to those that produce the foodstuffs to feed our population.

   I also want to take this opportunity to commend the FBI for its ongoing investigations into this, and other, terrorism incidents in Pennsylvania. They are doing a great job with the existing statutes but would also benefit if the statutes are strengthened.

4) We recommend that a significant portion of any federal appropriation for addressing the bird flu threat be devoted to field work and adequately preparing the agricultural community to prevent a bird flu outbreak or to promptly contain it

should it ever occur. While using a portion of the funds to stockpile serum to be used in the event of an outbreak is appropriate, funding should also be available to provide proper biosecurity signage at farms, compile a database of all animal and poultry operations, and properly equip our diagnostic laboratories to meet the challenges of a bird flu outbreak.

While I am on the subject of bird flu, I want to take the time to remind this committee and the general public that it is important to recognize why bird flu incidences have occurred in Asia and Eastern Europe. Most of the countries which have reported outbreaks have humans and birds and animals cohabiting in the same house or in very close proximity. These birds and animals run loose and, if they are carrying a virus, they are spreading it rapidly throughout the house of its owners and throughout the neighborhood.

One of the many advantages of modern confinement animal and poultry operations is that the animals are confined in environmentally controlled barns thereby minimizing the risk for infection. Confinement also eliminates the potential for wild birds and animals to spread the disease from flock to flock since the domesticated animals are isolated from the wild ones.

We submit that those that oppose our modern methods of animal production do not fully understand the multiple benefits of this approach.

5) One of the most frequent disruptions of poultry production in Pennsylvania is the problem associated with Live Bird Markets. These markets in some of our major cities are reservoirs for disease which can be easily spread back to the farms that are supplying the birds because cages and transport vehicles are not properly cleaned and disinfected. We encourage the USDA to expand its efforts to inspect and monitor these live bird markets in our major metropolitan areas.

6) Finally, while most of our comments and recommendations are centered on our animal and poultry agriculture, one remaining issue that also requires increased attention, is our agri-chemical industry. This industry, which PennAg also represents, provides the pesticides and fertilizers to those in productions agriculture. Agri-chemical facilities are also subject to the threat of terrorism. In this instance also, we believe that the federal statutes should be strengthened so as to discourage terrorism at these facilities which are so important to the success of production agriculture. Penalties should be increased to minimize or eliminate the potential for terrorism against this industry.

PennAg appreciates the opportunity to offer these comments on behalf of our 600 plus member companies who are the support and service sector of the number one industry in Pennsylvania, agriculture. I would be happy to respond to your questions.

Thank you.

Pennsylvania Field Hearing
of the
Senate Committee on Agriculture, Nutrition and Forestry
Subcommittee on Research, Nutrition and General Legislation
Pennsylvania Farm Show
Harrisburg, PA
January 9, 2006

Submitted by: Lew Gardner, Galeton, Pennsylvania

Title:

Bio-Security: Is the U.S. Agriculture Industry Prepared to Protect Itself?

Introduction:

Thank you for providing me with this opportunity to address the Senate Agriculture Committee today regarding the agriculture industry's ability to address bio-security issues.

My name is Lew Gardner. I own and operate a family dairy farm in Galeton, Pennsylvania. Today, I'm here in my capacity as a Pennsylvania dairy farmer. I also serve as the Chair of the Northeast Council of Dairy Farmers of America (DFA). DFA's Northeast Council has 1,700 dairy farmers who produce approximately 3 billion pounds of milk annually. In addition to my role as DFA's Northeast Council Chair, I am a Vice- Chair on DFA's Corporate Board. I am also a member of the Board of Directors of the National Milk Producers Federation and the National Dairy Board.

I want to thank Senator Santorum for his interest in bio-security and for holding this hearing today in conjunction with the Pennsylvania Farm Show in Harrisburg. We must continue to raise the awareness of this important topic in the interest of maintaining the health and safety of our nation's livestock industry and the consumer's confidence in our food supply.

The safety and security of the milk supply is of the utmost importance to the U.S. dairy industry and to DFA. Through its own initiatives, and in conjunction with industry partners, DFA has been involved in a cooperative effort to address challenges in protecting the nation's milk supply from a variety of threats.

For the past three years, the dairy industry has been working closely with the U.S. Department of Homeland Security, the Food and Drug Administration, the U.S. Department of Agriculture and other government departments to further safeguard the milk supply from potential threats.

Biosecurity is an important aspect of insuring a safe food supply and protecting animal health. Keeping food products wholesome and of the highest quality is important for the health and welfare of consumers. The ability to provide safe food helps to insure consumer demand, therefore helping to insure the profitability of farmers. Healthy and productive animals benefit farmers and enhance farm profitability and viability as well.

DFA's Farm to Market Bio-Security Activities:

As part of this ongoing work, DFA has taken a number of measures to secure the milk supply throughout the production and processing chain.

- Over the past three years, DFA led an effort between dairy farmers and processors to create and implement a milk tanker seal program. The new standards -- adopted by the vast majority of the industry -- call for milk tankers to be sealed following cleaning at their delivery point and again after loading milk from farms. The seals are only to be broken once the milk is delivered at a processing plant, and any unauthorized efforts to remove the seal on the tanker will result in the milk being scrutinized for tampering.

- During the same period, DFA endorsed emergency warehouse and milking room procedures as well as emergency guidelines for an "on farm" field staff protocol in the event of a herd health crisis.

- Over the past three years DFA redeveloped its crisis management plans, created a special crisis management web site with unilateral tools for team use, and has formed crisis teams at all levels (from farm to plant) in an effort to be pro-active, trained and aware of potential security threats from farm to processor to customer.

- DFA's manufacturing plants are registered to the Food & Drug Administration's database, and have completed Operational Risk Management (ORM) studies. Plants have secured entry systems, restricted access to the plant floor, implemented restricted access visitor and contractor programs and revised receiving and shipping programs. Packaging operations are automated and enclosed. Product packaging is generally tamper-evident.

- DFA has endorsed and published information about dairy farmers' use of bio security guidelines as recommended by the National Milk Producers Federation and the International Dairy Foods Association.

- DFA is in the process of launching an aggressive dairy farmer awareness and education program called "MySecurity." The goal of DFA's information campaign is to make dairy farmer members more aware of security and safety steps they can use "on farm" to increase consumer confidence about the security of farms, bulk tanks and milk supply from farm to market.

- <u>Components of the initial MySecurity rollout kits</u>

    - Multilingual "Restricted Access" signs are being delivered to members' farms

    - Multilingual "DFA Guide to Farm Biosecurity" wall charts

- Individual farm "Emergency Action Plan" templates with phone contact information for all agencies/personnel

- "Keep America's Food and Agriculture Safe "USDA bulletins (2) that highlight farm biosecurity issues

- Access to web based information through a secure connection for members to receive updates of farm security related material

All members will receive quarterly newsletters and security updates through email/fax/regular mail.

➢ Primary responsibility for DFA's interactions with the food security community as well as coordination of the development, and implementation of its internal "farm to fork" security initiative(s) has been placed in a newly established position at its corporate headquarters location.

➢ DFA is a member of INFRAGARD, a partnership of industry and the FBI to share information through secure email sites and promote participation in local, state, and national chapters of the organization to interact with those involved in the national security effort. A more agriculture specific group, to be known as AGGARD, is being created and DFA will participate in that effort.

➢ DFA is an active participant in the National Milk Producers Federation Foot and Mouth Disease research working group. This effort involves meetings of the various agencies that will be involved in oversight of any intentional/natural animal/milk related disease outbreak and is equally focused on plans to insure continuing milk marketing avenues in the event of animal quarantine operations.

➢ DFA has created a signage program and specific wall chart/handout material to be used by members that have visitors to their farming operations.

➢ DFA has created a global positioning satellite (GPS) location database for members, haulers, DFA plants, and customer plants along with corresponding milk production/capacity information so that electronically maps and/or area quarantine statistics are available.

Industry Initiatives:

Dairy farmers closely monitor their herds to ensure their health and well-being. Many producers have implemented recommended measures for security and bio-security. These measures include but are not limited to the following recommendations:

➢ Sealing and locking milk bulk storage tanks

- Having a combined entrance and exit for the farm with the farm name and contact information posted
- Restricting access to areas where milk, feed, farm chemicals and animal health products are stored
- Restricting and monitoring access to buildings and grounds through the use of fences, security lighting, alarms and appropriate lighting and signage
- Keeping records on visitors and deliveries and restricting their access to appropriate areas on the farm
- Maintaining a healthy herd through proper feeding, housing and care
- Purchasing healthy animals and quarantining them before they are exposed to the herd
- Require livestock haulers and dealers to disinfect equipment before entering farm complex
- Carefully screening job applicants, including background checks
- Create a crisis response plan

These voluntary recommendations are in the best interest of the dairy industry but one must use common sense before considering making them mandatory. 95 % of Pennsylvania's dairy farms have less than 100 cows. 25% of Pennsylvania's farms are owned and operated by Amish and Mennonite farmers, many of whom do not use electricity. In addition to working through the practicality of many of these recommendations, there will be a considerable cost to implement them. This cost cannot be borne entirely by dairy farmers.

The implementation of these guidelines and practices must be uniformly applied on a national basis and be in concert with current state regulations. For example, one could not require that milkhouses and bulk tanks be locked in Pennsylvania, while the U.S. Public Health requires them to be open and accessible for random U. S. Public Health rating inspections. If a rating officer was denied access to a farm facility, the farm would score a zero, possibly resulting in a bulk tank unit losing its right to ship milk under the PMO.

Dairy product safety and security doesn't conclude with good bio-security practices on the farm. Dairy industry partners continue to work closely with Federal and State regulators to develop and implement good bio-security practices, to the extent that the U.S. milk supply and dairy products are among the safest and most highly regulated foods in the world. The U.S. dairy industry utilizes a number of practices and regulations to ensure dairy product safety in the food supply as follows:

- The PMO is a set of requirements that address milk production, hauling, pasteurization, product safety, equipment sanitation and labeling.

- Every tank load of milk entering dairy processing plants is strictly tested for animal drug residues and quality. Any tanker that tests positive for animal drug residues is disposed of immediately, never reaching the consumer.

- HACCP is a structured and scientific process used throughout the industry to help ensure food safety.

- New standards for sealing milk tankers have been implemented; ensuring that unauthorized opening of tankers is immediately evident.

- Proactive steps to increase awareness among employees about security measures at the farm and in processing facilities are being implemented.

- Dairy processing plants are continuing to implement entry security systems, employee screening programs and restricted access to the plants operations.

- Packaging operations are automated, enclosed and secure.

The industry remains committed to security and safety and regularly evaluates the milk supply chain to keep America's milk supply safe.

Additional Work Remains:

While much has already been done, there remains a great deal left to do:

- The U.S. Animal Identification Plan needs to be implemented as soon as possible. Several breed associations, National DHIA and the National Milk Producers Federation have joined forces calling for the implementation of the plan prior to USDA recommendations. IDairy is calling upon Congress and USDA to provide the necessary resources and leadership to accomplish a timely implementation of this plan. A mandatory animal ID program provides a critical link in insuring healthy livestock and maintaining consumer confidence in our nation's food supply.

- Dairy farmers must recognize the need for the implementation of better bio-security practices on their farms. They must understand that these changes are in their best interest and that government resources will be available in part to support them. It would be helpful to dairy farmers if materials are developed that would help elevate the level of awareness at the farm level.

- National agricultural bio-security regulations must be carefully coordinated by one federal agency in order to insure that all federal regulatory partners are working together in a clear and concise manner. It's imperative that federal and state regulators understand the agriculture industry and are sensitive to providing practical solutions to improve bio-security protocols in the dairy industry. Providing spokesperson training to those who will be in positions of decision making and media interactions will help in the dissemination of a clear and concise message to consumers farmers and processors in the event of a bio-security related incident.

- The resources necessary to implement additional bio-security practices on the farm must be supported with Federal and State funds. These new regulations will be expensive to implement and the costs cannot be borne entirely by the farmers

- Federal Regulatory Agencies should support and recognize State regulations that are already in place and achieving the necessary results.

Conclusion:

The U.S. dairy industry has been working together for the past several years to improve bio-security practices on the farm and in processing plants. Keeping the dairy industry, it's livestock and products safe and secure has always been important to dairy farmers. Today's environment requires farmers, haulers and processors to re-evaluate their production practices to prevent the intentional introduction of livestock diseases and product tampering.

If dairy farmers want to maintain their markets and consumer confidence, we have to make farms and plants as secure as possible. We will be successful if we continue to coordinate our efforts and identify the resources necessary to successfully develop and implement good bio-security practices for the U.S. dairy industry.

The United States Senate Committee
on Agriculture, Nutrition and Forestry
Subcommittee on Research, Nutrition,
and General legislation

Presented by:
Keith Masser, President
Pennsylvania Cooperative Potato Growers Inc.

January 9, 2006

On behalf of production agriculture, The Pennsylvania Cooperative Potato Growers, and the National Potato Council, I want to thank your distinguished Committee for seeking input from grass roots organizations on agriculture bio-security.

I am the eighth generation in my family to own and operate a farming operation in western Schuylkill County Pennsylvania. After receiving an Agricultural Engineering degree from the Pennsylvania State University, I worked for Procter and Gamble as a Project Engineer at a plant in Mehoopany, PA that converts hard wood trees into personal care paper products. In 1976, I joined and eventually purchased my families farming operation, Sterman Masser Inc. Today this company farms 3600 acres of potatoes, hay and small grain crops. It also packages and distributes over 4600 truck loads annually of potatoes, onions, sweet potatoes and sweet corn to super market chain stores and food service providers in the Mid-Atlantic region.

In 2004, I developed and built a $12 million potato processing facility, Keystone Potato Products LLC. This company converts raw potatoes into dehydrated potato products using landfill gas as its energy source for the steam used in the dehydration process.

The third company I operate as president is the Pennsylvania Co-Operative Potato Growers located in Harrisburg PA. Organized in 1922 this Co-Operative is a non-profit organization helping growers market their potatoes.

Nationally, I recently was the president of the National Potato Council an organization that provides a voice for the 6000 growers making up the US potato industry on trade, environmental and legislative issues. I also served as chairman of the US Potato Board a national promotional organization funded by all US potato growers to create demand for potatoes and potato products.

Regionally, I chair the agricultural advisory committee to the Susquehanna River Basin Commission. This committee provides input on agricultural issues relating to the mission of the SRBC.

Statewide, I serve on the board of directors of the Pennsylvania Vegetable growers association and on the PA Potato Research Program.

Locally, I chair the Hubley Township Board of Supervisors since 1977, I chair the Schuylkill County Farmland Preservation Board and serve on the Schuylkill-Carbon County PA Farm Bureau board of directors.

Both Sterman Masser, Inc. and Keystone Potato Products use independent third party food safety programs to audit our food safety and bio-security practices. We are held accountable for traceability and chain of custody. With Keystone Potato Products being a food processing plant, customers such as ATEECO Foods and Campbell Soups require a HACAAP plan and an independent third party food safety audit program. With Sterman Masser Inc. being a farming and fresh potato packaging company, the American Institute of Baking is used. SMI uses the program voluntarily as the food safety audit programs are not yet required by the retail chain stores being serviced. Many growers and shippers of fruits and vegetables in the Mid-Atlantic region are not involved in such programs and in the near future will be forced to do so or loose their markets. Some state organizations such as the North Carolina Department of Agriculture have developed food safety audit programs and sell services to local grower shippers. It may be worth having an association such as the PA Coop contract with a Food Safety Audit provider and sell the services to it's members. This is being done by some sales organizations who are marketing national brands such as Green Giant and Dole Fresh. The costs of these programs ultimately need to be passed on to the consumer.

The role of government needs to be in the area of protecting the accidental or forceful introduction of pests, standardizing the crop protection chemicals allowed for imported foods and maintain those standards to equate to those used in the US. It needs to maintain programs such as the Golden Nematode program in NY to contain pests that have already been introduced into this country. We need to continue to fund research to help eradicate existing agricultural pests. This country's agricultural industry needs to enhance protection at it's borders from accidental and forceful introduction of agricultural pests. The Federal Government needs to keep emphasis on maintaining strong leadership at APHIS and provide additional training on plant and animal health issues to the Homeland Security Agency.

I believe the role of private industry should be to fund those demands created by market forces and let the level of standards be raised by competition and free enterprise. Government imposed programs need to be funded by government resources.

Thank you for allowing me to comment on this important topic.

Keith Masser

DOCUMENTS SUBMITTED FOR THE RECORD

JANUARY 9, 2006

Pennsylvania Farm Bureau
510 S. 31st Street • P.O. Box 8736 • Camp Hill, PA 17001-8736 • (717) 761-2740 •
www.pfb.com

Pennsylvania Field Hearing of the Senate Committee on Agriculture, Nutrition, and Forestry Subcommittee on Research, Nutrition, and General Legislation, Pennsylvania Farm Show Harrisburg, PA

January 9, 2006

Bio-Security: Is U.S. Agriculture Prepared to Protect Itself?

Statement by Pennsylvania Farm Bureau President Carl T. Shaffer:

On behalf of the Pennsylvania Farm Bureau's 37,500 members, I want to thank Senator Santorum for the leadership in organizing this event and for the invitation to offer comment. My participation today at the annual convention of the American Farm Bureau Federation prevents me from being with you at this hearing.

Pennsylvania's farm families believe that the protection of our nation's food supply needs to be a significant priority for our country and Commonwealth. Farmers have made great strides in implementing on-farm bio-security measures and providing information to animal health professionals to ward off potential bio-security threats. Our state government and research institutions have excelled in providing factual and timely alerts on disease outbreaks.

However, more can be done to expand and strengthen bio-security on Pennsylvania farms. First, producer education must be continued to ensure bio-security measures are implemented effectively at the farm level. Secondly, Congress should enact stronger penalties for those who purposely trespass on animal confinement facilities, which greatly increases the risk of spreading disease. Thirdly, we need to better inform consumers to avoid needless concerns that affect their decisions regarding the consumption of our food products.

These and other points are expanded upon in the written testimony provided for the hearing. Thank you again for convening this important event.

Pennsylvania Farm Bureau Testimony on Bio-security
by Carl T. Shaffer, President

On behalf of our 37,500 farm family members, thank you for the opportunity to testify today. The protection of our nation's food supply is a top priority for Pennsylvania farmers, the commonwealth and the nation. Senator Santorum's decision to hold this hearing today reflects the importance of this issue. The future well-being of our nation is vitally dependent upon continuation of a safe and readily available food supply. Recent acts of foreign and domestic terrorism, and serious outbreaks of disease in foreign lands, highlight the vulnerability of our food supply if effective measures to protect our farms and farm products are not implemented.

Farmers practice bio-security daily on their farming operations. Some bio-security measures are as simple as knowing the agribusiness professionals who stop by their farms regularly while also realizing when an unknown individual is visiting unannounced. Other bio-security measures are standard procedure such as showering before entering a sow farrowing barn or wearing suits and boots before entering a laying hen facility. Dairy facilities have foot-washing stations for their employees and hoof-washing stations for the cows to protect from diseases. In fact, bio-security is becoming common practice in Pennsylvania. Many industry professionals realize its importance and know not to enter more than one facility in a given day to not spread disease.

However, diseases and threats to our food supply change daily. Therefore, our protocols and technology must adapt. More can be done to improve bio-security on Pennsylvania farms. First, producer education must be continued to ensure bio-security measures are implemented effectively at the farm level. Producer education is a continuing process. Organizations such as Farm Bureau and other producer groups, as well as Cooperative Extension, currently provide education and information about the importance of bio-security on farms.

Unfortunately, purposeful threats to our industry are much more difficult to deal with. For example, in the poultry industry there is a recent concern of intentional or accidental contamination of flocks through exposure of unauthorized people entering poultry facilities. Secondly, Pennsylvania Farm Bureau urges Congress to enact stronger penalties for those who purposely trespass on animal confinement facilities, which greatly increases the risk of spreading disease. The recent illegal entry of an animal rights group into a Pennsylvania poultry operation with cameras that documented their own trespass on a local television news cast illustrates how such activities can dangerously breaching biosecurity measures. Furthermore, such behavior should be admonished by the news media rather than be used as a news account that advances the group's radical agenda.

Thirdly, we need to better inform consumers to avoid needless concerns that affect their decisions regarding the consumption of our food products. Farmers provide safe, quality products to our consumers and that message needs to be heard and understood. This can be accomplished through stronger relationships with media and industry. For example, the recent concern over the introduction of avian influenza (AI) into the U.S. has caused consumer concern.

The U.S. has many safeguards in place to detect and eradicate AI. No poultry infected or suspect to be infected with AI are processed for sale as raw meat in the United States. The United States imports very little poultry, which significantly reduces the possibility of exposing our industry to Bird Flu. Furthermore, virtually all of the chicken and turkey sold in the U.S. is produced here. Our consumers in the United States have little chance of coming across meat from an AI infected bird. Also, there is no danger of acquiring AI from properly cooked food.

The Pennsylvania Farm Bureau supports an overall greater commitment of resources from all levels of government to help us protect our food supply from acts of bioterrorism. While we believe Pennsylvania is on course in maintaining and further developing such safeguards, we need to continue the awareness and funding levels to ensure that our system can meet current and future demands. Thank you for the opportunity to present testimony on behalf of the Pennsylvania Farm Bureau.

QUESTIONS AND ANSWERS

JANUARY 9, 2006

Hearing to review the Benefits and Future Developments in Agriculture and Food Biotechnology
June 14, 2005
SENATOR TOM HARKIN
Questions for the record

Question for FDA, Dr. Brackett

The accidental or inadvertent presence in the food supply of materials from biotech crop varieties unapproved for food can cause major disruptions in the marketplace, putting branded food companies and export markets for US farmers at risk. Given the nature of growing and handling crops, there is a real likelihood that small and unintended levels of unapproved biotech varieties will be detected at times in the food supply in the future.

In August of 2002, the White House Office of Science and Technology Policy published a Federal Register Notice directing EPA, FDA and USDA to address concerns associated with low levels of unapproved varieties, primarily from field tests of new crops.

In November of 2004, FDA published a proposed guidance, "Draft Guidance for Industry: Recommendations for the Early Food Safety Evaluation of New Non-Pesticidal Proteins Produced by New Plant Varieties Intended for Food Use". This guidance partially addresses the possibility that material from a new plant variety intended for food use might inadvertently enter the food supply before its sponsor has fully consulted with FDA. This is a voluntary early food safety assessment process for some new biotech crops. It covers crops intended for food or feed, but does not cover biotech plant varieties developed for non-food uses of crops, such as plant-made pharmaceuticals. This proposed guidance, which is in addition to FDA's premarket consultation process, only partially addresses the issues of adventitious presence of unapproved new plant varieties in the food supply.

What additional steps has FDA taken to fulfill the 2002 White House directive?

Are there tolerance levels set for the adventitious presence of approved, or unapproved, biotech crops in non-biotech crops?

Should crops that produce pharmaceutical and industrial products have to meet the same food safety criteria as food crops, even though they will not be marketed as food, as a precaution in the event of adventitious presence of these crops in non-biotech crops?

As the variety and quantity of biotech crops increase, will FDA work with EPA to determine the risks of cross-pollination from unapproved biotech crops to non-biotech and to establish a tolerance or threshold for cross-pollination in these instances?

FDA's ANSWER:

In August, 2002, the Executive Office of the President, Office of Science and Technology Policy (OSTP) published a Notice in the *Federal Register* (67 FR 50578) which proposed coordinated actions by FDA, EPA and USDA aimed at strengthening controls over field trials to address the potential of material from field trials inadvertently getting into food or feed.

As part of this OSTP initiative, on November 24, 2004, FDA issued a draft guidance document entitled, "Guidance for Industry: Recommendations for the Early Food Safety Evaluation of New Non-Pesticidal Proteins Produced by New Plant Varieties Intended for Food Use." This draft guidance outlines procedures to address the possible intermittent, low-level presence in food and feed of new non-pesticidal proteins from biotechnology-derived crops under development for food or feed use but that have not gone through FDA's premarket consultation process. The comment period closed on January 24, 2005, and the more than 3000 comments received are under review.

The OSTP notice applied only to crops intended for food or feed use, and the FDA draft guidance also only applies to crops intended for use as food or feed. FDA is considering whether a similar guidance would be appropriate for new crop varieties of species ordinarily used for food that have been modified to produce pharmaceutical or industrial substances. However, such crops currently represent only a tiny percentage of new crop varieties under development and are grown under stringent confinement conditions established under an Animal and Plant Health Inspection Service (APHIS) permit. Of course, if material from such crops were to be found in the food supply, FDA would take appropriate regulatory action consistent with its statutory authorities.

In order to be sold legally as food or feed in the United States, crops that produce and contain pesticidal substances must be covered by a tolerance or an exemption from tolerance issued by EPA. FDA does not believe that there is currently a need or basis to establish tolerances for biotechnology-derived crops other than those that produce and contain pesticidal substances. Through an interagency process, we are working with USDA and EPA to continue to ensure that all biotechnology-derived crops are grown under such conditions as are necessary to maintain the safety of the food and feed supply.

Senator Tom Harkin

Question for USDA, Dr. Lambert and FDA, Dr. Brackett

Although biotech crops have been widely adopted by farmers, few genetically engineered animals are ready for regulatory approval and commercialization.

One food application, Aquabounty's fast-growing biotech salmon, is pending approval at the FDA. Most of research on biotech animals for use as food and biological "factories" for producing human pharmaceuticals purified out of milk is being conducted by academics, government researchers, and small start-up companies. TransOva, a small company working on biotech animals, is located in Iowa.

There is currently no formal government policy outlining how biotech animals will be regulated, so companies seeking to commercialize biotech animals (or find investment capital) have no clear pathway to the market. Statutory authorities at FDA, USDA and EPA could all be invoked to cover some aspects of biotech animals. However, the federal government has not publicly indicated how it will use those authorities, by themselves or in combination with any of their other existing authorities, to regulate genetically engineered animals.

The White House Office of Science and Technology Policy was coordinating an interagency review of this issue, but its efforts seem to have slowed down to a stop. In general, there has been a relatively clear, understandable process in place to evaluate the safety of new crops for human consumption and the environment. To my knowledge, there is no such process in place for new biotech animals.

Is there any formal policy for how the federal government intends to review genetically engineered animals? And if not, can you tell me how and when you will develop that policy?

FDA Answer:

Currently, FDA is using the investigational new animal drug provisions of the Food, Drug, and Cosmetic Act to oversee the safe development and use of genetically engineered animals. An interagency process for determining how best to regulate genetically engineered animals and their products is underway, and FDA is an active participant in that process.

APEX
TP
9780160773716
76901953-3